Packaging Prototypes 2: Closures

CLOSURES

ANNE AND HENRY EMBLEM

Photography by John Suett

RotoVision

Packaging Prototypes 2

Published and distributed by RotoVision SA
Rue du Bugnon 7
CH–1299 Crans-Près-Céligny
Switzerland

RotoVision SA, Sales, Production & Editorial Office
Sheridan House, 112–116A Western Road
Hove, East Sussex, BN3 1DD, UK

Tel: +44 (0)1273 72 72 68
Fax: +44 (0)1273 72 72 69
Email: sales@rotovision.com
Website: www.rotovision.com

ISBN 2–88046–504–4

10 9 8 7 6 5 4 3 2 1

Design by James Campus

Photography by Andrew Perris and Paul Mattock at APM on pp. 40, 48,
52, 54, 60, 62, 64, 70, 72, 86, 88, 92, 98, 107, 112, 120, 124, 126,
127, 144 and 152

Template designs by Mick Bevan

Production and separations by ProVision Pte. Ltd., Singapore
Tel: +65 334 7720
Fax: +65 334 7721

ACKNOWLEDGEMENTS

In preparing the material for this book, the authors have relied heavily
on the co-operation of the many individuals and companies who have
very kindly provided samples and information. They would like to thank
the following, without whom the book would not have been possible:
Paul Aikens, Stuart Bailey, Jane Beresford, Becky Biever, Bob Bushby,
Steph Carter, Alan Chan, Debbie Clements, Janet Coan, Adrian Cudmore,
Bruce Drew, Guillaume Dussert, William Edwards, Tony Fewell, Midori
Ffrench, Richard Fullwood, Rebecca Hulme, Michelle James, Alan Kirby,
B. K. Laurance, Nikki Martin, Jorg Melchers, Hiroshi Nishio, Carl Norlin,
Kiyohiro Ochi, Robert Opie, Jenny Owen, Christine Parish, Ian Parker,
Damian Porter, Ivan Powell, Elaine Pretty, Jill Robertson, Tim Sheldon,
David Smith, Alex Souter, Peter Symons, Bill Taylor, Deborah Tilley, Alex
Tosh, Steve Turner, Mike Whiteley, Stephen Williams and Simon Winn;
plus the following: Andrew Streeter of Creative Packaging Solutions
Limited for drawing the authors' attention to a number of innovative
closures from Japan, Stephen Wilkins of the Child-Safe Packaging Group
for permission to reproduce some of the information in the section on
child-resistant packaging, and Bruce Drew of Aquasol Limited for
information on counterfeiting.

Special thanks also go to Deborah Dawton, Sophie Williams and Nicola
Fleet of Design Events, who collected together all the physical samples;
Erica Ffrench of RotoVision who painstakingly made sure all the
necessary information was available; Harold Oliver for his review of the
style and comprehension of the text, and Dixie Dean for his review of
the technical aspects of the work.

Finally, without Kate Noël-Paton's inspiration for such a work, it would
not have been created at all, so to Kate go the biggest thanks of all.

CONTENTS

PREFACE

The lid, cap, stopper or spout may not at first seem to be the creative zenith of a packaging designer's day. It may be perceived that cerebral and artistic energies are spent only in resolving the usual tensions between brand personality, information, aesthetics and functional constraints. The closure may seem a final, prosaic formality following the intense stimulation of creating the pack.

But to correct this misnomer, consider funnelling oil, for example, accurately through a well-designed pouring spout: somebody used their imagination in focussing on an everyday problem and found a way to resolve it conveniently, ergonomically and cheaply.

There have been great advances in recent years in designing convenience into pack closures, whether it be for ease of opening, dispensing or applying the packaged product. Convenience should obviously lie at the heart of any three-dimensional packaging design exercise and it is a relief to consumers everywhere that today this is increasingly the case. Despite this development, it is still common to find recently designed packaging that is surprisingly hard to get into: shampoo sachets and cardboard milk cartons are two ubiquitous and stubborn examples.

Inconvenience, however, can occasionally be a virtue. The tireless ritual involved in opening a bottle of champagne; tearing the foil, releasing the tiny loop of wire to expose the classic cork, which then must be eased away from the bottle neck with the thumbs; its final expulsion from its tight opening on a blast of froth, makes the irksome

design a pleasure for the consumer. Thus an ancient solution to a functional problem has come to embody the character and sense of anticipation associated with the product.

The closure is effectively the last bridge between the product and its user. It represents the final and probably the most influential point of interaction between manufacturer and market. It is the point at which the user first sees, feels, smells or tastes the packaged product. As such, it becomes one of the most fundamental aspects of packaging design to get right. Consumers quickly switch to other brands or products if a dispenser fails to dispense or a tear strip does not tear. For a major manufacturer, the cost of getting it wrong can make the investment in the right design pale into insignificance.

Packaging Prototypes 2 is an essential follow-up to the first *Packaging Prototypes* which focussed on carton design. It enables manufacturer and designer alike to appreciate the importance of seeing beyond the function of the closure. The designs featured in this excellent new book show us just how much value can be added to pack design by rigorous attention to this pivotal design element.

Anyone who either practises or studies physical packaging design will learn from the solutions that are presented here.

George Riddiford
Managing Partner
Brewer Riddiford Design Consultants

PSAG

Packaging design is a key element in giving a product its competitive edge in today's marketplace. Yet while there is a wealth of material published on the subject of graphic design and packaging materials, some of the more obvious elements have been neglected.

Closures are often seen as mundane, even though they play a vital role. Consumers would complain swiftly if products leaked, were difficult to open or close, or had tops which exploded under pressure. So the onus is on manufacturers and brand owners to provide them with what they want.

Their problem lies in keeping up with technological advances, both in terms of materials and techniques, not to mention consumer demands. In such areas, the Packaging Solutions Advice Group (PSAG) is in a unique position to be able to help. Set up in 1996 to provide designers and clients with up-to-the-minute information and practical advice about every aspect of their packaging needs, our members encompass all aspects of the UK packaging supply chain.

Given their calibre, designers who work with our team can be assured not just of being kept up to speed on the latest developments in materials, decoration and pack shaping techniques, but of having access to advice and practical tips on all aspects of packaging, from printing to design, production and implementation techniques.

That is why we were so keen to sponsor this book. It represents not just an opportunity to educate and inform designers and clients about a

unique part of the packaging mix, but enables us to set standards of quality and innovation in an area which is all too often ignored.

The following pages show the role of closures in clever packaging and good design. Remove all the graphics from a bottle of Coca-cola, HP Sauce or a bottle of Jif lemon, for example, and most of us would still recognise the products for what they are. The combination of pack shape and closure is unmistakable.

So what do the examples featured have in common? The panel of four industry experts who chose them were looking at four main criteria. The closures had to break new ground, be consumer friendly, provide added customer benefits and demonstrate technical innovation.

The result is a snapshot of today's market. What is remarkable is that not all of the devices featured are new, some are classics. The ring-pull, for instance, was brought out in the early '70s to hold in carbonated drinks yet it still pulls its weight some thirty years later, albeit with a few tweaks to its design.

In the following pages it rubs shoulders with the innovations of today, such as the clikpak™, a container for homeopathic pills. Designers have taken it on board that the pills should be taken straight from the tube, so the push mechanism dispatches the dosage directly into the cap which can be tipped into the mouth. Then there is the Spout Cap, a screw-cap which eliminates disasters when opening cardboard milk containers.

Such examples depend on a close relationship

8

The Packaging Development Company

Victor International Plastics Ltd.

RPC Containers Ltd.

Corus Packaging Plus

Decorative Sleeves Ltd.

Merck Ltd.

Gilchrist Bros. Ltd.

Tag Labels Ltd.

Packaging Solutions Advice Group (PSAG)

between the packaging industry, clients and designers. While the design process, and the skills needed for it, have become more sophisticated, so have the range of materials, techniques and markets. The rate of change is so swift that no company, in isolation, could keep pace.

The range of closures, meanwhile, points to changing attitudes towards the subject. The pack, its decoration, the closure and ease of handling are becoming much more of an integrated process. That is why we are bringing people from all parts of the packaging chain together, because companies are adopting a much more holistic approach to packaging development.

Our members possess knowledge of packaging materials, such as Corus (the former British Steel Tinplate), and RPC Containers; and of colouring and pigments, with Merck and Victor International Plastics. Decorative Sleeves and Tag Labels provide input on pack decoration; while the Packaging Development Company specialises in the management of packaging innovation and Gilchrist Bros. Ltd. in digital pre-press. We facilitate those designers who wish to collaborate with more than one of our team at a time, providing access to a range of leading-edge technologies in a unique and time-saving way.

Yet even with such a depth of understanding, the task of whittling the number of closures featured down to sixty-five was not an easy one. The result however, is a book which we hope will visually inspire and instruct. We have gone for quality rather than quantity, because otherwise the result

would have been too lengthy to be user-friendly.

This specific area of design represents a challenge to all those who work within it. For a closure to be a success these days it needs to meet a number of criteria. It must be consumer friendly, add as little cost to the manufacturing process as possible, provide a unique selling point, and be as easy to manufacture and make part of the product as possible. Those products shown in the following pages fit the bill. They also demonstrate what we, as a cross-industry group, stand for: bringing good design and innovative packaging to the supermarket shelf.

We know the size of the problem facing clients and designers. The questions they ask themselves are familiar to us, too: "How is it going to appeal to the consumer and how are they going to re-use, recycle or dispose of it?" "How is the pack going to be merchandised in the supermarket?" "How are we going to manufacture it?" "How are we going to get it through the distribution chain?"

There is no simple answer to any one of these questions, but the PSAG does have the depth of knowledge among its members to be able to provide a helping hand. We also hope that by sponsoring this publication, *Design Fundamentals' Packaging Prototypes 2*, we will provide designers with a source book for images, ideas and inspiration.

Packaging Solutions Advice Group (PSAG)

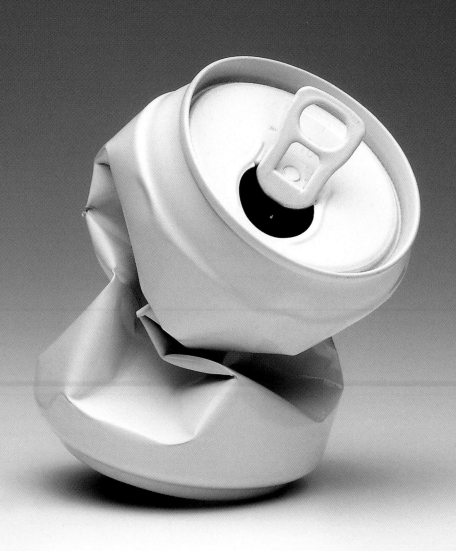

INTRODUCTION

Undoubtedly the most sought-after convenience feature on packed products is the ability to open and, if relevant, re-close the pack with ease. The closure is thus a fundamentally important feature of any pack; not only does it keep the contents inside the pack, it provides the means by which the user gains access to those contents. The type and style of pack closure directly influences the way in which the pack is opened and the product is used, as well as the safety of the pack and the user. One of the major objectives of this work is to provide designers with some practical guidelines to help inspire creative thinking in the area of pack closures.

Packaging today is such a common and everyday part of our lives we probably seldom stop to think about it at all, let alone consider why it is there and what roles it has to perform. Because it is so ubiquitous it is difficult to imagine life without packaging. Indeed, there are very few, if any, products which do not require packaging of one kind or another: food and drink, health and beauty products, hardware and electrical goods, are all prepacked; ready for us to select from the store shelf and carry home. Even fresh produce sold loose on market stalls, which at first glance may appear to be sufficiently well-equipped with nature's own packaging (bananas and oranges in their skins, for instance) needs packaging in the form of trays, crates or boxes, to get it from the field or orchard to the point of sale. Textiles and clothing, which we also frequently see on display without any obvious packaging around the goods, will nevertheless have been wrapped and packed at the factory, ready to survive the storage and handling stages en route to the point of sale.

The packaging around a product has to fulfil several important roles. While the relative importance of each role varies with the type of product, we need to understand these roles before we can even begin to consider what type and style of packaging will suit a particular end use. First, it has to contain the product. This means the pack must stop a liquid, paste or cream product from leaking. Product leakage invariably results in damaged goods and dissatisfied customers, and can potentially be a serious hazard if the product is a corrosive chemical such as some household cleaning liquids, or garden products. Containment also means keeping a number of products together, to be sold as one unit. Multipacks of biscuits or drinks, a gift set of toiletry items, or a pair of gloves or socks fall into this category. The customer expects these to be securely contained, with no likelihood of anything being missing when they open the pack at home.

Packaging has to protect the product against certain physical hazards to which it may be exposed throughout its useful life. This life span embraces the time from the point of production and packing, through the point of sale, through use of the product and finally, disposal of the packaging by the end user. However, the critical activities in this time, where 90% of all product damage happens,

11

are those between the end of the packaging line and the point of sale; operations which can be collectively called the distribution chain. It is the activities of storing, handling and transporting the product which are most likely to lead to damage.

Just some of the hazards in this chain are falls from conveyors, pallets or vehicles, or drops due to bad handling practices, which may result in denting, cracking, or outright breakage. All vehicles are subject to vibration, made worse by poor road surfaces, and this can lead to scuffing and scratching; it can even lead to screw-caps on bottles becoming loose. Goods stacked on pallets in warehouses will be subject to compression forces, which may lead to deformation or breakage, and goods stored at the wrong temperature, or in a damp atmosphere, can deteriorate rapidly. Another hazard, which will be dealt with later, is that of pilferage and tampering.

All of these hazards can cause damage to the product or the packaging, which can result in the item being left on the shelf; even a slightly dog-eared carton or a dusty or scuffed cap will cause the would-be purchaser to reject that product in favour of something more pristine. Today's sophisticated consumer is alerted by even the slightest evidence of damage, suspecting poor quality and poor value for money. Once disappointed, the consumer soon becomes critical of a certain product, or indeed, a whole range of products and looks elsewhere.

When it comes to food and drink products (which account for about 70% of all packaging used) not unnaturally, consumers expect them to remain wholesome and safe to eat or drink, right up to the end of their stated shelf life. Packaging has a major role to play here in preserving products, keeping them in a good and wholesome condition. The preservation role is also important for pharmaceuticals, toiletries and cosmetics, which must remain fit for their intended use and not deteriorate in storage or use.

There are many factors which can cause deterioration: growth of bacteria, exposure to air, and exposure to moisture, for example. The end result can be at best the unappetising one of soggy biscuits or cereals, or off-tasting crisps, and at worst it can be sickening or even deadly, for example food poisoning due to the E-coli or campylobacter organisms. The main principle of preservation is to effect some change to the condi-

tions which lead to the product's deterioration, and then use the packaging – sometimes in conjunction with low temperature storage – to maintain those conditions up to the point of use.

More than ever now consumers demand a wide range of different foods, available all year round and in a variety of pack sizes to suit the needs of different family units. The gradual disappearance of the corner shop, coupled with the inability and/or reluctance of consumers to shop for food more often than is absolutely necessary, means that the reliance on packaging as a means of helping to extend shelf life has become ever more important. Of course, there is nothing new in using packaging in this way: the development of sterilising food by heating it in a sealed metal container (the forerunner of today's 'tin' can) goes back to Napoleon and his need to feed his army during long periods away from home. The sterilisation process kills the harmful bacteria and the hermetically sealed can then acts to keep air and moisture away from the product until the pack is opened. 'Hermetic' means there is no exchange at all between the inside and the outside of the pack; in practice there are few truly hermetic packs.

The difference between today and the nineteenth century is that in addition to normal canning (in the UK we buy some eight billion cans of food per year) we now have a choice of other preservation methods available to us, which allow us to preserve a wide range of products without changing the taste or texture to any noticeable degree. Fresh milk, for example, is routinely pasteurised (heated very briefly) during the packaging process, and, provided the packaging remains intact, the product will then keep for several days in the fridge.

One big growth area of food preservation has been in chilled and frozen convenience foods, where ready-prepared dishes are packed in handy-sized portions. They are usually packed in shallow trays with sealed lids, plus an outer sleeve or carton. The food can be heated in, and even eaten from, the tray, thus bringing a range of both speciality and basic cooked dishes to consumers who are increasingly demanding 'good' food, but who lack the skills or the time required for its preparation. The packaging here is an integral part of the product, the two working together to produce a total system of food preservation, presentation, preparation and consumption.

Another good example of the integration of product and packaging is the development of Modified Atmosphere Packaging (MAP), where the atmosphere surrounding the product is artificially changed, usually to reduce the amount of deterioration-causing oxygen. Nitrogen and carbon dioxide, used singly or in combination, are commonly used. Once the optimum gas conditions are achieved, the packaging then continues the preservation role by providing a barrier to the loss of the modifying gas and/or gain of air from the surroundings. This calls for both high gas barrier packaging materials, and, crucially, a very high level of integrity in the pack seal, which must remain gas-tight until the pack is opened. This can have implications for ease of opening the pack, especially for the elderly or infirm, this will be discussed later.

Typical products preserved by MAP are fresh and cooked meats, cheese, salads and vegetables, all of which also require chilled distribution and storage to maintain their shelf life. MAP products which can be kept at ambient temperature include yeast, nuts, pasta and part-baked bread.

Vacuum packaging, where the surrounding atmosphere is modified to the extent where the air is totally removed, can be thought of as a variation of MAP. The difference is that due to the vacuum formed by the expulsion of air, the packaging collapses onto the product and conforms very closely to it, whereas in 'conventional' MAP there is always a space around the product for the modifying gas. The same packaging demands apply: i.e. high gas barrier materials and integrity of seal; but vacuum packaging is only suitable for products which can withstand the close conformity of the packaging without becoming damaged. For example: ground coffee and peanuts are good candidates for vacuum packaging, while crisps, cream cakes or delicate salads such as lettuce are not.

In contrast to the requirements mentioned so far, where there is a high level of emphasis on the need for packaging with good barrier properties, certain products have a specific requirement for packaging with a very low moisture barrier. Fruit and vegetables continue to respire and produce moisture after picking and if they were placed in tightly sealed and impermeable packaging, the moist conditions would soon be just right for the growth of moulds and thus deterioration of the products. The same is true of morning goods and crispy bread, which are usually packed warm, hence the widespread use of perforated film bags.

Packaging has a role to play in providing information about the product, as even just a cursory glance at any range of packed products will show. This includes information which is required by law: the name of the product, its weight or volume, ingredients, how to store, how to use, any safety warnings etc. Consumers are far more litigious nowadays than at any time in the past, and coupled with a large amount of governmental legislation, this has meant a growth in the amount of information which manufacturers are legally required to give about a product. The packaging is invariably the only medium available to convey information and there are strict rules concerning such factors as size of type and position on the pack.

There is also a need for information which helps to move the product through the distribution chain, to make sure the right product reaches the right point of sale, and is sold at the right price. The use of bar codes, which capture a whole raft of information in a small printed area, is now commonplace; they not only identify the goods, but they are also used to control stock levels, helping to make sure the retailer or wholesaler doesn't sell out of a particular line. Another example in this category is the growing use of small identity labels on individual fruits such as apples, so that the supermarket check-out operator can quickly distinguish a Granny Smith from a Golden Delicious and key in the correct price code.

Most importantly, the graphic information on a pack, working in conjunction with the pack's colour, shape, and size, fulfils another fundamental role of packaging: that of promoting and selling the product. Many products are packed in opaque packaging, often for very practical reasons: avoiding colour fade, for example, in some bath and hair care products, or to reduce the risk of oxidation of fatty products due to exposure to light while on display. In some cases, opaque packaging may be preferable simply because the product is not particularly attractive to look at, which is the case for certain prepared foods and pharmaceutical preparations. It is, therefore, the packaging which takes on the role of attracting the would-be purchaser's attention as he or she scans the hundreds of goods displayed on the store shelves.

This attraction is achieved either by immediate recognition, much as one would recognise an old

13

friend, or by curiosity, where one is drawn to the unusual and stops to investigate further. Most shopping decisions are made by the former process, where we spot the product we want by its familiar colour or shape and reach out for it almost without thinking. But this familiarity must never be taken for granted by the marketeer, or the packaging designer: it is the result of a complex mix of factors, usually developed over many years into what we call a brand, often of inestimable value to the owning company. Equally importantly, it should not be abused by blatantly copying an existing pack colour, graphics or shape, to 'fool' the consumer into picking up a 'me too' product and trading on existing brand loyalty.

Arousing the shopper's curiosity is an equally complex task. The product and pack must combine first of all to draw attention, and then to hold that attention for just long enough for a decision to buy to be made. As we shall see in the next section, features such as pack functionality will be key here, but to be meaningful and effective, such features must be obvious to the beholder in that vital decision-making period.

Most of the discussion so far has been illustrated by using examples of what we call fast moving consumer goods (fmcg), produced in large numbers and sold in retail supermarkets and stores. However, many of the points are equally applicable to other goods: domestic appliances, shoes and clothing, which are normally displayed without their packaging. They also apply to prescription-only pharmaceutical products, and to chemicals and food used in manufacturing processes. While the selling function of packaging is unimportant for these products, the containment, protection, preservation and information functions remain vital and must be considered as part of the design of the total pack system. Similarly, the transit packaging for all types of goods must also meet these functions. A packaging designer may not view a corrugated case as a particularly exciting or challenging piece of packaging, but if its task is to contain 48 packets of biscuits and it collapses in the warehouse, the loss of product can be costly. This loss will be many times more if, say, the product is an expensive pharmaceutical product.

How does packaging work?

Once it has fulfilled its role of attracting the consumer's attention and setting up the expectations which have encouraged purchase of the product, packaging then has a vital role to play in assisting the product to meet those expectations. In addition to containing, protecting and preserving, all of which will almost certainly be taken for granted, packaging works in practical terms by delivering the product into its intended market, and to the end user, in such a way that it is convenient and safe to use.

It is this convenience aspect of packaging with which consumers become most familiar, albeit often without thinking about it until a problem arises. The pack which can be confidently and safely picked up, put down and stored in its intended place, without falling over and spilling its contents attracts no attention or comment. It is only when the pack fails in some way that the consumer notices its inadequacies, and may be driven to seek out an alternative brand or product on the next shopping trip.

Convenience in terms of being able to open and if required reclose a pack is of prime importance here and this is very clearly a function of the design of the pack closure. It is therefore appropriate at this stage to consider the following question: What is a closure?

THE IMPORTANCE OF THE PACK CLOSURE

In its most basic sense, a closure may be defined as any device used to close a pack such that the pack fulfils its roles of containing, protecting and preserving, to the extent demanded by the product. This includes everything from the simple twist tie on a plastic bag to the most sophisticated multi-part injection moulded cap. Broadly speaking, it also includes the way in which a sachet is sealed together along its edges or the way in which a carton is closed by tuck-in flaps. However, this work will concentrate mainly on showing examples where the closure is a separate component which is applied to a container and held there, usually by some mechanical process.

Other than the open-topped trays used extensively in the distribution of bread and other products into the retail market, we can say that every pack has a closure. We can also say that the effectiveness of a pack is only as good as the effective-

14

ness of its closure. For example: as a packaging material glass provides a total barrier to the passage of moisture; thus, a glass jar is capable of preserving a moisture sensitive product for a very long time. The extent to which this capability will be realised is entirely dependent on the efficiency of the closure used for the jar. If this is out-of-specification with respect to its dimensions, moulded in a highly permeable material, or badly applied, the preservation function of the pack is compromised accordingly. While the requirements of this basic definition of a closure must never be overlooked, as has been said, a pack's closure has a direct influence on the efficient and safe use of the product. Good packaging design recognises this degree of influence and takes due consideration of market demands and expectations. To explore these demands and expectations in more depth the following four key, and in many ways conflicting, areas affecting choice and design of closures have been selected for review: tamper evidence; child resistance; ease of use and packaging safety and dispensing and measuring features.

FOUR KEY AREAS OF CONSIDERATION AFFECTING CLOSURE DESIGN AND SELECTION

Tamper evidence

The need for tamper evidence is a late twentieth century phenomenon, almost certainly sparked off in the USA in the 1980s when Tylenol pain-killing tablets were deliberately contaminated with cyanide, resulting in several deaths. The reaction which reverberated very rapidly throughout the USA and Europe was a review by manufacturers and sellers, to question just how resistant their packs would be to the wilful tamperer. While this review may have been confined initially to the pharmaceutical sector, it soon spread to food and drink products, as responsible companies in these sectors began to consider the consequences of wilful tampering.

The introduction, by someone with a grudge against a company or sector of society, of a foreign body into a product at best results in an expensive product recall operation. The almost inevitable loss of consumer confidence following the adverse publicity may also affect the company's image and spread across their entire brand, with disastrous financial consequences. If packaging was not already high up on the agenda in the boardroom, it soon became so, as companies recognised their vulnerability in this area.

The answer to the question 'can this pack be tampered with?' will, of course, always be 'yes', to some degree. Just as one can never totally protect one's property against the determined thief, no pack can ever be totally resistant to the determined tamperer. Hence the 'tamper-proof' pack, for all practical purposes, is non-existent. What manufacturers and sellers do demand, however, is packaging which provides a deterrent to the opportunist tamperer, either because of its complexity, or because signs of attempts to open it can be readily noticed and recognised, thus making the tampering ineffective. It is this latter requirement of tamper evidence which has become an important consideration in packaging design.

DEFINING TAMPER-EVIDENT PACKAGING

Tamper evidence is defined by the American Food and Drug Administration as:

"Having an indicator or barrier to entry which, if breached or missing, can reasonably be expected to provide visible evidence to consumers that tampering has occurred."

Unlike child-resistant packaging, for which there are internationally recognised standards setting out a procedure for determining if a closure can be deemed 'Child-Resistant' and to which manufacturers can work, there is as yet no equivalent standard for tamper evidence. Thus, companies do not have specific standards to meet in order to make this claim for their pack. However, in the UK a new Code of Practice was issued in September 1998 by INCPEN – the Industry Council of Packaging and the Environment – and endorsed by Government and the Institute of Packaging, calling for packaging to be designed to minimise both pilferage and tampering. This code states: "If there is a health risk to the consumer as a result of deliberate contamina-

15

tion of the contents, it is recommended that tamper-evident packaging should be considered".

EXAMPLES OF TAMPER-EVIDENT PACKS

Examples of how packaging is used to minimise tampering and to provide tamper evidence now abound in the food, drink and pharmaceutical sectors, as a review of the designs in this work will show (see Table of Designs). The basic principle of most tamper-evident packs is that they incorporate some device into the closure, which provides a means of making it obvious that it has been opened. To be effective, this relies heavily on the consumer recognising that tampering has taken

place, and rejecting that pack, drawing it to the attention of the seller or manufacturer. Tamper-evident features, therefore, need to be very obvious and stand out clearly from the 'normal' pack.

One of the oldest examples, pre-dating the Tylenol incident by a couple of decades, is the metal cap used on bottles of spirits. Its design and use was brought about by the need for anti-counterfeit measures (of which more later) to discourage unscrupulous operators from refilling an empty bottle with sub-standard product, then passing it off as the real thing. It is known as the Roll-on Pilfer-Proof (ROPP) closure, its name testifying to its age, as it has already been said that it is now accepted that pilfer- or tamper-proof packs are an unrealistic goal. It is made by stamping and drawing a shell out of a printed and lacquered sheet of aluminium and introducing a line of perforations around the circumference. A sealing wad is also fitted. At this stage the cap has no threads. On the filling line, the shell is dropped over the neck of the filled bottle and held in place while a special tool rolls the relatively soft aluminium to conform to the threads of the bottle and then folds it around a ring on the bottle neck.

When the cap is unscrewed, the skirt portion below the perforations falls beneath the neck ring and becomes trapped on the bottle, thus giving visible evidence that the bottle has been opened.

The ROPP metal closure is generally confined to bottles made of highly rigid materials, glass being a perfect example, where the neck can withstand the rolling process without the threads becoming

distorted. When it comes to less rigid materials such as the many plastic bottles used for mineral waters and soft drinks, variations on the same theme using plastic tamper-evident caps have been developed. These plastic caps are made to a high level of dimensional accuracy by injection moulding. The threads are moulded in – to meet standard neck finish dimensions and tolerances – and the skirt is held in place by perforations. The cap is screwed in place to a pre-determined torque value on the filling line, and when unscrewed by the consumer, the skirt portion breaks away and stays on the bottle. For maximum 'evidence' the bottle should be designed so that the skirt falls down the neck section, leaving a very clear gap between it and the remaining cap. This readily draws attention to the fact that it has been opened.

Other plastic tamper-evident caps require the user first of all to remove a band of material by tearing, before the cap can be taken off the bottle. Common examples of this include the caps used on plastic milk bottles, glass and plastic bottles for cooking oil, and plastic tubs for vitamins and other health supplements. Similar principles are used on a range of plastic boxes, tubs and pots for biscuits, salads, ice creams, etc. Here, the lid incorporates a locking tab and it is only when this tab is broken away that the lid can be removed from the container.

The metal cap with the pop-up safety button is probably one of the most well-recognised tamper-evident closures around, due to its widespread use on jams and preserves, meat and fish pastes, and baby foods. It is a feature of these products that they are either filled very hot, or cooked in the jar, and on cooling a partial vacuum is created in the headspace inside the jar, pulling the metal cap down into a concave shape. Opening the pack releases this partial vacuum and the metal springs back up, giving an audible 'pop'. The vacuum is permanently lost and when the cap is put back on the jar, the now domed button is easy to see and can be pressed down, leaving both visible and audible evidence of opening. Most manufacturers print information on the cap to draw the consumer's attention to this feature, e.g. "Reject if centre button can be depressed". Reflecting the sensitivity of such products, baby foods go one step further and have a tear-off plastic band as well as the pop-up button feature.

If meeting requirements for tamper evidence can be combined with fulfilling some other important role of packaging, so much the better, and a good example of this is the growing use of shrink sleeves. These are cylindrical sleeves of plastic film, which can be printed with high quality graphics, giving all-round decoration to a bottle or jar and thus enhancing the selling function of the pack. If clear films are used the sleeves can be printed on the inside surface; this gives a high gloss finish and a scuff-resistant print. Extending the length of the sleeve so that it also covers the cap adds the tamper evidence, as the cap cannot be removed without tearing a portion of the sleeve. On the filling line the sleeves are placed loosely over the pack which is then passed through a heated tunnel, causing the plastic film to shrink snugly onto it and take up the contours of the total pack. As a decoration technique sleeving is especially effective for packs with compound curves and unusual shapes, where the application of labels would be limited.

To be an effective tamper-evident feature, the portion of the sleeve which has to be removed to gain access to the cap should be printed in such a way that its removal would be very obvious to the eye. To maintain the decoration, branding and information requirements of the pack, the sleeve should be perforated along the bottle or cap interface, so that the top portion tears off cleanly, leaving the rest intact.

A variation of shrink sleeves is the simple shrink band seen on jars of spreads and which needs to be torn away before the jar can be opened. Again, printing these makes them stand out so that the consumer is likely to notice if they are missing.

Another example of a multi-function approach is the clear film overwrap used on cartons of tea, cigarettes, chocolates, perfumes etc., which has to be removed before the carton can be opened. The film can be specified to provide a moisture barrier to keep the product in optimum condition, e.g. to stop tobacco from drying out. Or it can provide an odour barrier, either to stop harmful odours getting in (e.g. chocolate, which is highly susceptible to picking up 'off' smells from neighbouring products) or to stop volatile flavours from escaping and thus changing the product's taste (e.g. tea). The film also enhances the pack's appearance and gives a luxury image to the product, hence its use in perfumery and skincare products.

For film overwrapping to work as a tamper-evident feature, the consumer has to be familiar with the fact that the product normally has a film around the pack, and notice if it is missing. Printing the film, and/or using a printed tear tape gives additional security, as it makes replication more difficult; the tear tape also provides a very neat easy-open feature. There are two basic styles of wrapping used, the parcel wrap, where the film is folded and tucked neatly around the carton, and the shrink film type, which uses thinner film and conforms very closely to the carton. The latter is more difficult to remove intact and replace, and hence possibly more secure.

Sealing a container across the opening by means of a diaphragm which has to be torn away to reach the product is another common way of providing tamper evidence; a simple example of this is the foil or film laminate lid used on single serving pots of yogurt and desserts. Where the pack is designed for multiple servings an additional cap is needed as a reseal feature. The induction inner seals used on glass jars for dried instant coffee are another good example of diaphragm seals, as are the numerous paperboard-based containers used for baby milk formula, custard powder, sugar, flour and drinking chocolate.

For these dry products, the diaphragms can be anything from a plastic coated paper to a heavy-duty aluminium foil, dependent on the degree of moisture barrier required. Some are fitted with easy-open pull-tabs, while others have to be pierced with a knife. As tamper-evident features, these seals all have one thing in common: because of the additional cap, any unlawful attempt at entry will not be noticed until the pack has been taken home and opened. The same can be said of the foil laminate diaphragms sealed across the necks of plastic bottles for milk, sauces and syrups. In many of these the diaphragm is also a key contributor to the containment function of the pack and once it is removed the pack must be stored upright to prevent leakage.

An interesting variation of the diaphragm concept is the thin membrane of metal sometimes seen across the open end of aluminium tubes. This has to be pierced, usually by means of a pointed feature moulded into the cap, in order to squeeze out the product. The membrane is formed as an integral part of the impact extrusion process of making the tube.

Applying a self-adhesive label over the tuck-in flaps of a carton, or to the point where the bottom edge of a cap meets the jar or bottle, is another way of approaching tamper evidence. It is sometimes used as a 'quick fix' solution, perhaps in reaction to a publicised tampering incident, to give an immediate feeling of reassurance to the public and to fill the gap while something more substantial is developed. When highly decorated, these labels can also enhance the product image, as seen on jars of preserves with 'saddle' labels applied to the top surface of the cap and wiped down the sides, giving the appearance of an 'official seal'.

To be effective, the adhesive on a label used as a tamper-evident measure needs to form a very strong, permanent bond to the pack, ensuring that it cannot be easily peeled away. The label material itself should be relatively weak and tear or distort irrevocably when the pack is opened, providing clear evidence of tampering. However, to be truly

effective, it must not be possible to replicate the label in any way, to prevent dishonest traders from simply reapplying a new label. This moves us into an area closely related to tamper evidence, that of anti-counterfeit packaging.

18

ANTI-COUNTERFEIT PACKAGING

Counterfeiting means fraudulently imitating a product and is perhaps best known in the forging of banknotes. For our purposes it means taking the outward appearances, i.e. the packaging, of a well-known and reputable brand which is known to command a high price, and substituting a cheaper, sub-standard product. This can mean either refilling the actual packaging, or copying the design of the branded packaging and reproducing it for filling with the inferior product. Both types of counterfeiting are particularly relevant for high value products such as alcoholic spirits, top-of-the-range 'designer' perfumes and, of most concern, pharmaceuticals. The World Health Organisation estimates that some 10% of all pharmaceutical products sold worldwide are counterfeit. Estimates in other sectors are not readily available.

The measures available to manufacturers to protect themselves against this crime include security of packaging materials, especially of printed items, and security of finished stock throughout the distribution chain. But there are ways in which innovative packaging design can make a significant contribution here.

For counterfeiting to work, the fake product must be indistinguishable from the real thing, to the untrained eye. This relies on the consumer recognising enough elements of the 'brand' offered to believe it is genuine, and the easiest way for the fraudster to achieve this is to make use of empty packaging from the original product. One measure already being used in the spirits sector to counter this is to insert a special fitment into the neck of the bottle which effectively prevents the bottle from being refilled, while at the same time it does not impede the normal flow of liquid out of the bottle. Attempts to remove the fitment result in it breaking, leaving part of it trapped in the neck of the bottle.

Another way to discourage the counterfeiting of a branded design is to incorporate into the pack a feature which cannot be easily copied. Holograms similar to those used on credit cards can be printed on labels. A new development is a patented security device called PULSLINE®. This is a cleverly designed label using a combination of substrates, adhesives and banknote threads, which can be applied to the pack so that it has to be removed to gain access to the contents. The ultra-strong adhesive means that the label cannot be removed without itself being destroyed in the process. The security thread prevents fabrication of a replacement label and provides a means of validating the packed product both at the end of the packaging line and at the point of sale. The thread can be made so that it is visible to the naked eye, or only visible under UV light, which can provide a covert means of checking authenticity. This development may be the ultimate in product security, providing both proof of authenticity and tamper evidence.

Of course, there are a vast number of products which may be regarded as inherently tamper-evident, because the pack function is destroyed on opening. Examples which come to mind are the tin and aluminium cans used for a wide range of food, soft drinks and beers, as referred to in the first section; as well as the crimped on 'crown corks' used on beer bottles, which have to be levered off. In the pharmaceutical area, the glass ampoules used for injectable medicines have to be broken open before they can be used. Similarly, push-through blister packs used for tablets – where the aluminium foil has to be broken to get the tablet out of its formed plastic pocket – come into this category, as do pouches and sachets which have to be torn or cut to open.

However, it must be remembered that none of these would present the determined, malicious tamperer with an insuperable task. All of the examples given in this review concentrate on how the pack closure, i.e. the normal method of entry to the product, can be made tamper-evident. The skilled tamperer is unlikely to be bothered by such convention and it should be borne in mind during the packaging design and development process that packs may be violated at points other than the closure.

Despite the examples mentioned so far, the requirement for tamper evidence is not confined to consumer products. Industrial products require protection also; for example: food ingredients being delivered by tankers in bulk. Responsible producers, aware of their accountability (not to mention legal requirement) to deliver safe and wholesome products to their customers, use uniquely numbered security seals or electronic systems, to provide this protection.

Finally, the need for consumer vigilance must not be under-estimated, which comes back to the importance of the consumer recognising that something is different about a pack. This aspect must be considered at the packaging design stage.

Child Resistance

The natural inquisitiveness of young children means a tendency to play and experiment with almost anything, including any number of normal household items beyond the territory of the toy box. If this is a packet of chocolate biscuits pulled out of a cupboard, the result is probably no worse than sticky fingers and perhaps a stained carpet. However, if it is a bottle of caustic dishwashing liquid or paint remover, the consequences can be very serious and even life-threatening. The same can be said in the case of pharmaceutical tablets or liquids. The estimated figure for cases of ingestion of medicines alone, by children in the UK, for example, is around 45,000 per year.

This worryingly high figure is published by The Child-Safe Packaging Group in the UK, a body set up in 1995 with an objective to: "promote the specification, consideration and success of child-resistant packaging for all products where ingestion or other contact could prove seriously distressing to a child".

This objective raises two interesting points for the packaging designer, both of which require further discussion. First, the need to understand what constitutes 'child-resistant packaging' and second, the need to consider the product being packed, in the light of its potential effects on children, as opposed to adults, to decide if child-resistant packaging is required.

DEFINING 'CHILD-RESISTANT PACKAGING'

A good definition of child resistance is: 'Packaging that is difficult for children to open within a reasonable period but represents no difficulty for adults to use properly.'

It is important to stress that just as a pack can never be made tamper-proof, we can never guarantee that a pack will be child-proof. This is not a realistic goal and the use of child-resistant packaging must not be seen as a substitute for taking normal preventative measures to ensure children do not have access to potentially dangerous products, e.g. by storing in out-of-reach or locked cupboards.

There are two international standards (ISOs) for determining child resistance. EN 28317 has been in place since 1985 and applies to recloseable packs, i.e. capped bottles or jars containing several doses of a product, where there is a need to repeatedly open and reclose the pack. EN 862 was introduced in 1997 and applies to single use packs such as the push-through blister packs commonly used for tablets, where the pack is destroyed once the product is removed. However, EN 862 has not been adopted by the Pharmaceutical Industry.

Both standards set out a protocol for testing packs to determine child resistance, and call for the tests to be carried out using children and adults. The children must be within a particular age group (42 to 51 months) and at least 85% of them must be unable to open the pack within a prescribed period (five minutes for recloseables, three minutes for single use packs). The group must then be given a non-verbal demonstration of how to open the pack, after which at least 80% must fail to open it, again within the prescribed period of time. The number of children required for the test is 200. For the group of adults, which must number 100, at least 90% must be able to open (and if relevant reclose) the packs. The tests must be conducted by an authorised test house (of which there exists only one in the UK, for example).

Thus there are laid-down standards for deciding on whether or not a pack is child-resistant, although, given that humans vary in intelligence and dexterity, test procedures that involve groups of children are not ideal. However, it is encouraging that since the introduction of testing, the incidence of ingestion of hazardous products by children has fallen. While the degree to which this is due to greater public awareness, rather than the use of child-resistant packaging, is not known definitively, there is little doubt that this has made a positive contribution. (This information is quoted from EN 862:1997).

PRODUCTS WHICH REQUIRE CHILD-RESISTANT PACKAGING

The interest in and requirement for child-resistant packaging in the UK came along with the growth of consumer legislation in the 1970s, and the introduction of the Medicines (Child Safety) Regs., 1975. Initially, these Regulations made it mandatory to use child-resistant closures only on children's Aspirin and Paracetamol, but they were subsequently extended to include adult versions of these drugs. The now widespread use of child-resistant closures across virtually all other pharmaceutical products is due largely to self-regulation by the pharmaceutical industry and the Royal Pharmaceutical Society, as well as manufacturers and retailers.

Outside of the pharmaceutical area, products which fall within the Chemicals (Hazard Information and Packaging for Supply) Regulations, 1994 (known as CHIP2) and subsequent amendments, are legally required to be fitted with child-resistant closures if supplied in recloseable packs. These products include substances which are officially classified as 'toxic' or 'corrosive', and preparations containing more than a given percentage of certain organic solvents. However, CRCs are often used even when a product does not come within these Regulations, as manufacturers and retailers assess the potential risks of their products, and opt for the responsible and safe approach, rather than risk a company-damaging law suit. Thus the use of CRCs continues to grow, especially on household and garden chemicals, again largely as a result of self-regulation by industry.

In considering whether or not child-resistant packaging is required for a product, it is important to note that a seemingly harmless product to an adult, may cause totally different reactions in a child. Some mouthwashes, for example, have a fairly high alcohol content. This will have little or no effect on an adult, who understands that the product is swilled around the mouth and spat out. The child, on the other hand, may swallow the coloured liquid and the effect of the alcohol content on the relatively low body weight could be quite marked. This is an example of a product which clearly is not classified as toxic or corrosive, is not a household or garden chemical, and yet has the potential to cause distress to a child.

Thus it is vital to consider each product on its own and try to assess its effects. This applies not just to the effects of ingestion, but also on skin, eyes etc. The length of time to which a child may be exposed to the product must also be considered. If an adult receives a splash of bleach on the face, the immediate reaction is rapid drenching under cold water, which probably limits the damage to a slight and temporary reddening of the skin. The same splash on a young child, who does not have that same reaction, will mean a much longer contact time and greater potential for lasting damage.

In summary, a good guiding principle for deciding on whether or not child-resistant packaging is required is to be found in the objective of the Child-Safety Packaging Group, quoted at the start of this review, i.e. products where ingestion or contact could cause serious distress to a young child.

EXAMPLES OF CHILD-RESISTANT CLOSURES

There are several different approaches to designing recloseable child-resistant closures (CRCs) for bottles and jars, almost all of them requiring two simultaneous actions to be performed. The principle is that this requirement will present a challenge to a child's cognition but will not present a barrier to an adult. Most have some form of instructions moulded in or printed on to the closure, using diagrams or written matter, which will inform the adult, but have little or no meaning to a child within the prescribed age group. Examples of such closures can be seen in the marketplace and several designs are shown in this work (see those asterisked on pages 32–34). Just a few examples will be reviewed here.

The 'push down and turn' closure usually consists of two parts: an inner screw-threaded part and an outer shell. To remove it from a bottle the outer shell has to be pushed down to engage with the inner part, and only when this is achieved can the cap be disengaged from the screw-thread on the bottle, by turning anticlockwise. Turning without first pushing down causes the outer shell to rotate, but it cannot be removed – no matter how many times it is turned, or how vigorous the turning. Additionally, a very useful feature is that this action of turning causes a loud audible 'click' which may act as a warning to parents or those nearby. This type of closure is very widely used on glass or plastic bottles of tablets, including prescription medicines dispensed by the pharmacist.

Another type used on screw-threaded containers is the 'squeeze and turn' cap, where the outside of the cap has to be squeezed at two designated points, which only then allows it to be unscrewed. The 'squeeze points' are usually subtly indicated in the moulding and require moderate pressure to allow the cap to be unscrewed. Until they are located and squeezed the cap will not rotate. These closures are used on kettle descalers and kitchen and lavatory cleaning products.

Several styles of cap exist where two or more features have to be lined up before the cap can be removed. These are often the push-on, lever-off style and some require considerable force to lever off the cap, as well as good eyesight and manual dexterity to line up the relevant features. They are used for pharmaceutical tablets packed by the manufacturer and for pharmacist-dispensed prescription drugs. An interesting innovation in this field comes from Australian product design company, Toren Consulting in Sydney, who have developed a closure which allows release of a single tablet by lining up three dispensing apertures. A key improvement over the lever-off style caps is that the cap is not – and cannot be – removed to gain access to the tablets, removing the possibility of the cap being left off accidentally, or of the contents being spilled out.

Ease of use and packaging safety

Much of what has been said so far about pack closures has emphasised the importance of achieving and maintaining a high level of integrity,

i.e. a well-sealed pack to keep the product in prime condition throughout its expected life. In particular, the last two sections have reviewed ways in which packs can be closed, where there is a specific intention to confront a particular group of users with a pack which they will regard as 'difficult' to open. The objective is that the opportunist tamperer or the playful and curious child will meet sufficient resistance to lose interest in the task.

However, one of the big challenges facing the packaging designer is that these requirements of high seal integrity for the product, along with tamper evidence and child resistance, have a potential conflict with the demand for convenience in opening the pack and using the product. This conflict must be addressed, as it is not acceptable for the pack which offers protection at one level to present an insurmountable or dangerous challenge at another level, i.e. to the intended user.

PACKAGING ACCIDENTS

Packaging-related accidents are a serious issue, resulting in more than 60,000 people being treated in hospital each year in the UK alone. This does not include the figure of 45,000 accidents involving medicines and children, given at the beginning of the review of child-resistant packaging. As there is no logical reason why accident statistics will be significantly different in the rest of Europe and North America, it can be said that the problem of packaging safety and ease of use is real.

PACK OPENABILITY

Approximately 40% of reported accidents are related to plastic and glass bottles and jars. Problems of openability are of particular concern. Injuries can arise due to the bottle neck breaking off, or the use of an unsuitable opening tool, such as a knife to lever off a cap. Metal caps with the pop-up safety button, mentioned already for their positive role in offering tamper evidence, are particularly difficult to unscrew due to the partial vacuum inside, and with increasing age comes decreasing strength and manual dexterity. The problem is exacerbated if the cap is smooth edged and circular, as this means that it is difficult to grip firmly. Design features to improve openability include knurling the edge of the cap, and using square, hexagonal etc. cap shapes, all of which will help to make the cap easier to hold without the hand slipping.

The Roll-on Pilfer-Proof metal cap and its many plastic moulded variants were also mentioned as common and effective tamper-evident devices. However, if for some reason the skirt portion does not 'break' at the perforations, it may be impossible to remove the cap without resorting to the use of mechanical assistance, often in the form of a knife as the nearest implement to hand at the time. If the result is a physical injury requiring hospital attention, this is

another incidence of a packaging accident. It may well be true that the user should have been more careful in wielding the knife, but it is also true that if the cap had performed as required, there would have been no need for the knife in the first place. Making sure the cap does perform to standard is largely a matter of quality assurance, but the designer would be well-advised to consider this aspect when designing the 'break ring' or similar type closures.

Canned foods, identified as an excellent means of providing long-term shelf life and inherent tamper evidence, are also implicated in packaging accidents. Using the wrong opening tool is a partial cause, but the sharpness of the open edge even when the correct can-opener has been used still leaves a potential hazard. Easy-open can ends are starting to be used more and more, and are no longer reserved for the higher-priced end of the market, which is to be encouraged. Developments continue in this field and there is still scope for innovative ideas on ease of opening and safe removal of the product and handling of the can.

Another group of packs which has attracted a lot of criticism with respect to openability is those made of flexible packaging materials, examples being the all-plastic thermoformed packs used for bacon, cooked and raw meats, cheese and similar products. These materials are specifically designed to resist tearing and puncturing and to have a very high level of integrity in the pack seal to guarantee the product's shelf life; all of this can add up to a pack which is difficult to open. Fortunately, this is an area which has now received a lot of attention, and peelable systems which rule out the need for knives and scissors are becoming commonly available. Some of these also incorporate easy reseal

features, allowing the product to be stored in its original wrapping until used up – which is of environmental benefit, as it avoids the need for 'clingfilm' or aluminium foil kitchen wrap.

CONSIDERATION OF THE USER – THE AGEING CONSUMER

The examples given so far all serve to highlight the need to consider both how the pack is going to be used, and by whom, when designing suitable packaging closures. Reasonable conditions of use need to be considered, e.g. is the pack likely to be opened in the shower, when the hands are wet? If this is a bottle of shampoo or shower gel, the cap needs to be easy to open without having to read instructions. Or is the product meant to be eaten or used outdoors, when a special opening tool is unlikely to be to hand?

As far as the user is concerned, age, manual dexterity and general state of health are important considerations. Factors such as improved nutrition and standards of healthcare mean that people in the developed countries are now living longer than previous generations. However, ageing brings a natural decline in strength and manual dexterity and even mild arthritis in the fingers will significantly heighten this decline. Eyesight is also highly likely to decline with age, and coupled with the natural human tendency not to bother to read instructions, means that these cannot be relied upon for the safe opening of a pack.

The needs of this sector of society cannot be ignored; almost 30% of the population of Western Europe are aged 50 or over, and their buying power is considerable, thanks to good pension planning and improved standards of living. The Centre for Applied Gerontology at the University of Birmingham in the UK has a stated principle which packaging designers should take on board: "Design for the young and you exclude the old; design for the old and you include the young."

Dispensing and measuring features

As well as providing a means of gaining access to the product, there are many opportunities in packaging design for using the closure to control the way a product is applied or used and to add to its effectiveness by controlling the amount dispensed. See the designs asterisked on pages 32–34 for just some of the ways in which this is currently done. Good dispensing and measuring features can make a significant contribution to the safe use of the pack and the product, and help to address some of the issues raised in the last section on ease of use and packaging safety. Alternatively, a closure that fails to dispense or measure efficiently will soon turn the consumer against the product, even though it is the pack which is malfunctioning.

EXAMPLES OF DISPENSING CLOSURES

There are many, many examples of dispensing closures, some traditional and some modern. One now traditional and relatively simple device for applying a viscous liquid product is the roll-on deodorant pack, where a plastic ball fits into the top of the container and acts as an applicator for the liquid product. The fit of the ball is crucial; too tight and it will not rotate and hence will not allow the product to be applied to the skin, too loose and the liquid will seep out and be applied too freely. The fit of the overcap onto the ball is also important: the cap must push the ball down so that it is seated on a special sealing ring inside the neck of the bottle. Failure to do this results in a leaking pack.

Other liquid dispensing devices include the anti-glug closures on motor oil, with extensible spouts for easy pouring into car engines. Similar features could help to reduce the hazards associated with pouring harmful garden chemicals. On the same principle, but with different applications and made to very different designs, pull-up spouts also lend themselves readily to use on a wide range of products, from washing-up liquid to drinks.

The latter are especially aimed at drinking 'on the run' and in the gym, and are used for mineral waters and sports drinks. The benefits are that the drink can be safely carried without spillage and when required, the closure can be opened with the teeth, with minimum disruption to the exercise regime.

Other products which benefit from some form of dispensing are the pre-wash liquids used as stain removers for laundry. The ability to apply the concentrated liquid directly to the soiled area, e.g. collars and cuffs, or socks, is a useful feature and bottles with trigger spray closures have long been used for this purpose. However, a range of Japanese products worthy of note in this sector includes one with a special nozzle for spreading the liquid, and another with a plastic brush moulded into the closure, allowing the soiled area to be scrubbed while the liquid is applied.

Caps which incorporate brushes for applying liquid and semi-liquid products also have useful applications. Examples can be seen on glue and on small pots of paint, which mean the product is ready to use, without requiring a separate brush, and this in turn means less cleaning up after the job is done. Similar applications are nail enamel and mascara, the latter also requiring an effective wiper mechanism to control the amount of product on the brush.

Solid products also have requirements for dispensing, perhaps the simplest example being the plastic sprinkler tops used on powdered products from talc to cocoa, and granular products from cake decorations to parmesan cheese. These often have differently-sized dispensing holes to allow for both sprinkling and pouring. Also, they usually have some means of closing off the dispensing holes to help prevent spillage and product deterioration once the pack is opened.

MEASURING OR METERING THE PRODUCT

Measuring features offer additional benefits to simply dispensing a product, giving a degree of control over the amount used. These vary from the simple measuring caps fitted to bottles of laundry liquids, where the accuracy of the dose is a function of the cap moulding and needs the attention of the user; to highly sophisticated metered dose systems used for inhalers and self-injectable products such as insulin. Design requirements for the latter are obviously more exacting than the former, as here the operation of the device should be such that the amount of drug dispensed is controlled independently of the particular individual, i.e. the device needs to be 'foolproof'. In general, the criticality of the dose determines the accuracy and complexity of the components used.

A wide range of different pumps and aerosols are used on perfumes, toiletries and cosmetics, delivering a set amount of product on each depression of an actuator. The benefits include: quick application, one-handed operation, and reduced likelihood of contamination. They are available for liquid and cream formulations, including highly viscous products.

Requirements for accurate dosage are important in many chemical and agricultural applications, and one innovative approach to this is the use of water soluble films (see page 104). The product is safely contained in a film sachet which is simply added to water, the film dissolves, and the correct dose is delivered accurately without having to measure out and handle a potentially hazardous substance. This type of packaging could be considered for any product, hazardous or otherwise, which is used in an aqueous environment. Other benefits include reduced likelihood of error in dosage, no contaminated packaging to be disposed of, and reduced wastage of expensive products.

The first consideration for any dispensing or measuring system has to be the nature of the product, and the consistency of its properties from batch to batch and throughout its expected life. Closure design has to take account of any normal variations in the product, and the consumer's natural desire to want to empty the pack as easily as possible. If the product is likely to harden on exposure to air, consideration should be given to some form of anti-clog device, for aesthetic and product performance reasons as well as consumer satisfaction. The clogged nozzle which effectively seals the product in the pack before it can all be used up is unlikely to lead to a repeat purchase.

Finally, a word about usage instructions for dispensing and measuring. It is a prime requirement that these be as simple, short and obvious as possible. Pictograms, where relevant, are an ideal alternative. If an explanation is becoming long and complicated, this might be a signal that the closure device itself is too complicated for practical application.

24

ASPECTS OF CLOSURE PRODUCTION

PRACTICAL LIMITATIONS

The factors limiting the design of a pack closure (and all other packaging components) include: firstly, the materials and processes used to make the closure in commercial quantities; secondly, the requirements of the packaging line where the closure is applied. In addition there are two overriding limitations which affect all packaging decisions: economic acceptability and the need for environmental responsibility.

Good, effective design cannot be achieved without an understanding of all of these factors. Of course, the packaging designer cannot be expected to have an intimate knowledge of all materials, processes and packaging line activities and so, to help in this regard, he or she is well-advised to forge close links with the people who do have this knowledge. Most likely, this will be the

packaging technologist, packaging development or materials specialist.

This section will deal with the manufacturing processes, briefly describing each one and then identifying the key design considerations. The material, packaging line, economic and environmental aspects will be briefly reviewed in subsequent sections. The main manufacturing processes with which we are concerned are injection moulding for plastics closures and various stamping and forming processes for metal closures. There is also some production of plastic closures using compression moulding. These processes cover the production of the vast majority of closures and closure systems, as a review of the designs in this work will show.

Production processes for plastic closures
Injection moulding: The injection moulding process

consists essentially of melting a prepared polymerised thermoplastic material and forcing a specific quantity into an accurately machined cavity enclosed in a block of metal known as a mould. Very high pressures are used, which call for heavy, steel moulds, held together by a high clamping force, all of which make injection moulding a high cost process. After the injection stage, the next stage is to cool the molten material and return it to a solidified state: this is achieved by means of a series of channels in the body of the mould, through which circulating fluid can flow. The body of the mould has small venting slots to allow trapped air to escape on closing. The mould is made in two parts (more for complex shapes) and after the cooling cycle it opens along the part-line and the cooled, moulded part is ejected. The mould then closes and more material is injected to start the process over again.

The time from injecting the molten material to ejection of the formed part and the mould being ready for the next injection is called the cycle time and this has a major influence on the total cost of the process. The rate of production can be dramatically increased by using multi-cavity moulds which produce anything up to sixty items per cycle, although this significantly increases the initial cost of the mould. This highlights the importance of knowing the volume requirements for the closure before the mould is commissioned: building a large multi-cavity mould means the number of units required must be sufficient to amortize the high capital cost over a reasonable period of time; building a mould which ultimately turns out to be too small means an unnecessarily high unit cost and the need to invest in a new mould very quickly.

The prepared material consists of the selected polymer, various additives to assist processing and improve performance and, if required, colourants. Additives which help the injection moulding process include heat stabilisers and mould release agents; those used to improve performance of the final part include antistatics, impact modifiers and UV absorbers. Colourants can be pigments or dyes: pigments are inorganic or organic compounds which are insoluble and are dispersed throughout the molten material; dyes are organic and are soluble in the polymer, which means they will produce transparent colours. The range of colours available is limited by the temperature stability of the available pigments or dyes, and the usage conditions required. Typically, inorganic pigments are the most resistant to the processing temperatures used and give the strongest colours, but if they contain the heavy metals cadmium, lead, mercury and hexavalent chromium, their use is now severely limited by legislation.

Concentrated pigments or dyes in carrier resins, known as masterbatches, are commonly used. These are prepared by specialist colour compounders and supplied in pellet form to the injection moulder for addition to the polymer at pre-determined levels. Universal masterbatches which are suitable for use in a variety of different polymers are available, but it cannot be assumed that these will result in the same final colour, irrespective of the polymer being used. Therefore if more than one material is being used for a total pack, for example a polypropylene cap on a high density polyethylene bottle, the colour achieved using the masterbatch in the two materials must be checked during development, and adjusted if necessary.

The amount of material forced into the mould is calculated taking into account the size of the part and the shrinkage of the polymer being used. All plastics shrink on cooling and each one has a specific shrinkage rate. This means that a mould must be designed at the outset for use with a specific material. Changing the material may require a different set of processing conditions and may not result in the required degree of dimensional accuracy in the part being made.

The site at which the molten plastic is injected into the mould is known as the injection point or gate, its location is an important consideration during the design stages of a closure. Its position will affect the even flow of material into the mould, as well as the aesthetics of the finished part. Ideally it should be as unobtrusive as possible and this may add to the complexity of the mould. Although unlikely to be required for simple closures, large or very complex designs may necessitate having more than one injection point, to give an even flow of material throughout the mould. Similarly, the location of the ejector pins which are essential to the smooth ejection of the closure from the mould must be considered at the design stage, as they invariably leave small marks which could detract from the appearance of the closure. As with all other surface defects, these are generally much more apparent on shiny surfaces and dark colours, than on sand-blasted or matt surfaces and pastel colours.

The molten material is fed into the mould cavities by means of channels or runners and for each cycle of the machine, the material in these channels becomes hard and is ejected with the closures, thus producing a matrix of scrap plastic each time. To avoid this, the channels can be heated and this is known as the hot runner system. The obvious advantage is of not using an excess of material, although the increased complexity of the mould is reflected in its increased cost.

The point at which the mould opens, known as the part line, also requires early consideration. If, as is highly likely, the closure has threads or undercuts it may be necessary for the mould to open at more than one location in order that it can be easily removed. This will also increase the complexity and therefore the cost of the mould.

One further design consideration for injection moulded closures is the need to avoid 'sink marks', which are small depressions on the closure surface. Good control over moulding conditions is important here, although sink marks can be difficult to eliminate completely, especially if the design is such that there is a big variation in the wall thickness throughout the closure – from thick to thin wall sections – which leads to different shrinkage rates. There are a number of different approaches to this problem.

Firstly, they can be disguised by surface effects such as the vertical ribs commonly seen on screw closures (which have the added benefit of making it easier for the consumer to grip the closure for opening). Secondly, they can be reduced considerably by the use of foam moulding, whereby an inert gas is dispersed throughout the molten material during the moulding process, causing the material to expand and fill the mould cavity. The part maintains its structure due to the foaming process, taking away the need for high pressure during cooling (which in turn means lower mould cost and thus a lower economic quantity). A third approach is to achieve the final effect using two components: a simple inner cap to fit to the container and a thin-walled outer cap moulded to the required shape. This is applied to the inner cap, usually as a push-fit. Finally, the use of a totally different process – compression moulding – offers another way of avoiding sink marks.

Compression moulding of plastic closures: Most plastics used in packaging application are what are called thermoplastic materials, referred to in the above section on injection moulding. This means they soften on heating and harden on cooling, and the heating and cooling processes can, up to a point, be repeated without the polymer breaking down. This

is a very useful property of thermoplastics, as it means that scrap mouldings can be re-melted and reprocessed.

A second group of plastics is known as thermosets: these are polymers which, during the heating process become cross-linked, with very strong chemical bonds, and on cooling harden into rigid structures. However, any attempt to reheat the material results in permanent breakdown and it cannot be reprocessed. The first synthetic plastic to be made, phenol formaldehyde (PF), is an example of a thermoset material and, along with urea formaldehyde (UF), is used for some closures in the pharmaceutical and cosmetics industries. PF closures are always dark in colour while UF is used for white and pastel shades.

Like injection moulding, the compression moulding process requires specially-built moulds which are usually multi-cavity. The prepared material is in powder form and is metered into the mould and then compressed and heated, which brings about the cross-linking and forms the required shape. The advantage over injection moulding is that thick wall sections are readily achievable without surface sink marks. Another advantage is that both PF and UF have very high levels of chemical resistance, especially to perfumes and solvents. Also, the finished closure is always much heavier than the injection moulded equivalent, which adds to the high quality image desirable in the cosmetics sector. The process is usually slower than injection moulding and the unit cost is often higher, although the lower tooling cost makes the process useful for modest quantities, which again explains its application in the cosmetics sector.

Production processes for metal closures

Metal closures are made by stamping the required shape from a sheet and then subjecting this 'blank' to several forming processes. Tin-plated steel, tin-free steel and aluminium are all used. Metal closures are invariably printed with a design or an all-over coloured lacquer and this is applied to the flat sheets prior to stamping, usually by the offset litho process followed by stoving of the inks and lacquers at high temperature. While this results in good adhesion of the ink or lacquer to the base metal, it means that the choice of colours used in the printing inks is limited to those which will remain stable during this stoving process. Printed designs impose a greater degree of accuracy on the stamping process than plain colours, due to the need to accurately register the stamping tool to the design – the more complex the design, the more critical this part of the operation.

Forming the metal blanks into closures is achieved using multi-stage tooling to produce the required shape, with, where required, the correct thread formed and the bottom edges rolled or trimmed neatly. Aluminium, being more ductile than steel, is easier to form although as a softer metal it may be less robust in use.

The tooling to produce metal closures is usually complex and expensive; this means in practice that most are made to 'standard' designs where the tooling is owned by the closure manufacturer. The designer chooses from a range of standard designs and introduces originality via colour and print. While this can also be applicable to plastic injection moulded closures, it is also true to say that there is greater opportunity to have a custom-made shape produced in plastic than in metal.

Closure wadding

A fundamental aspect affecting the efficiency of a closure is the way in which it seals at the interface between the closure and container and this usually requires the use of a pliable material to fill any gaps between the two components. Plastic injection moulded closures can be designed to fulfil this role unaided – provided the material is sufficiently soft and of accurate dimensions vis-à-vis the container – and so-called wadless closures are now common.

Wadless closures often seal on the inside of the container neck, known as a bore seal, and their advantage is that as they do not require the addition of a separate wad, this removes one stage in the manufacturing process. They also eliminate any potential problems due to wads falling out during transit of the closures to the filling line. However, for a bore seal to be truly effective there must be good control over the accuracy of the dimensions of the inside of the container's neck. This require-ment may have implications for the manufacturing process used for the container, and consequently its cost.

Metal closures (and in practice many plastic closures) require the addition of a wad of material to conform to the sealing area of the container. The choice of material for this wad is determined by the flexibility required and a range of different types is available. The simplest is paperboard faced with a plastic film (waxed cork was the original wadding material). The paperboard or cork gives the degree of compression required and the plastic film forms

the seal and provides the required barrier to mois-ture or gases. Expanded plastic materials are also used. Whatever material is chosen, there is always a requirement to accurately cut out the wad and fix it into the closure so that it does not fall out. This can be achieved using adhesives, or via an interference fit inside the closure.

Wads are also achieved by using liquid plastic materials known as Plastisols, which are 'flowed in' to the closure during its manufacture and cannot be separated. They are soft, rubbery materials and conform readily to the neck of the container.

PACKAGING MATERIALS

The choice of material used for a closure (and wad) is determined by the demands of the product, the expected life of the pack, the need for recloseability and the desired product image. Product/pack compatibility is paramount and there are legislative requirements to be observed, especially for plastics, where there are limits for the level of migration of certain substances from the packaging into food.

Performance properties of materials, such as the barrier offered to the passage of moisture, oxygen, carbon dioxide and light are known and are the remit of the packaging technologist who will have access to the required data. This data can be used as a first guide to material selection, ruling out the obviously unsuitable and guiding the designer. The thickness of aluminium and steel used for closures can be regarded as impermeable, but the permeability properties of plastics, including those used for wads, will vary dependent on the selected plastic and should be checked, taking into account the required shelf life of the product and how it will be stored, handled and used.

If a closure is designed for repeated use, the flexibility of the plastic material will be important, especially in the case of 'flip-top' caps moulded in one piece with an integral or 'live' hinge. Un-doubtedly the best commonly-used plastic for this application is polypropylene. For all screw-threaded plastic caps the material's resistance to creep (which, among other effects, is responsible for loss of torque and unscrewing over time) is a factor for consideration and again polypropylene offers good performance in this regard.

The image created by the product in the market-place will be influenced by the choice of packaging material and decoration. Where a luxurious image is required an anodised aluminium overcap on a perfume bottle will be preferable to a plastic alternative. Similarly, high gloss surfaces may be desirable although, as these are likely to emphasise the appearance of surface defects, a compromise may be needed.

The role of economics in closure design will always be a significant part of the decision-making process. The designer needs to be aware of both the total packed product cost allowed and the budget available for the development process, including any costs of market research, promotional activity and advertising. He or she also needs to be familiar with the costing process, to appreciate which factors have the most significant influence and which can be changed with minimal effect.

The total packed product cost will include all purchased ingredients and packaging materials, all labour directly and indirectly associated with the production, packaging, storage, handling and distribution of the product, and running costs such as energy, fuel and services. It will also include a proportion of the depreciation costs of the packaging line, the company buildings and other fixed assets. The bought-in cost of a closure will be influenced mainly by the cost of the material selected, the production rate of the closure and the cost of any custom-built tooling which will have to be amortized over the number of units purchased. The cost on the production line will depend on the ease and speed of application to the container and the rejection or wastage rate; modern packaging lines demand accurate performance often at high speed with little or no manual intervention. All of these costs can be directly influenced by the closure design, which can therefore make the difference between a product being economically feasible or not.

The relative contribution of packaging material and component costs to the total cost of a packed product will vary enormously within different product sectors. Basic food and drink products are very keenly priced on retail sale, and are usually produced in large quantities with low margins. At the other end of the scale, high-priced perfumery and skincare products are likely to be less margin-sensitive. On the face of it, this latter category may seem to offer greater scope for creativity, although this is not necessarily true: the challenge presented by the need for maximum machine efficiency which usually goes hand-in-hand with simplicity of components, while at the same time delivering

innovation into a fast-moving marketplace, certainly requires creativity and must not be underestimated.

There is a need to consider the environmental implications of everything we do today and packaging design and selection is no exception. Manufacturers have always investigated material reductions for cost reasons, and environmental legislation is now forcing them to minimize packaging and to provide evidence that they have done so.

Designers have an important part to play here, in influencing material weights and minimizing waste, which will affect use of resources. Also, once a pack has fulfilled its designated roles and the product is consumed, disposal of the pack must follow and the chosen method of disposal must be designed into the pack and specified at the outset.

The process of packaging development starts by assessing the demands of the product and its intended market, drawing up a list of critical values which must be met for success. If a product is fragile, the degree of fragility needs to be understood; if it is moisture sensitive, the extent of this sensitivity needs to be determined. If the target market for the product is school-age children, the requirements of this sector need to be defined as closely as possible and not simply stated in vague terms. The hazards to which the packed product is likely to be exposed throughout its entire life cycle need to be defined: if the product is being shipped to tropical countries, the climatic conditions prevailing therein must be specified. Other demands will be imposed by packaging machinery, legislation, competitors' activities and time constraints, as well as economics as already mentioned.

All of this information is compiled into a Packaging Brief, a portfolio of requirements which is used as a checklist when searching for suitable materials and assessing candidate designs. The earlier in the product development process that the packaging brief is developed, the better, and ideally product and pack development should be carried out in tandem and not sequentially. Similarly, packaging solutions should be developed at this stage not just for the primary pack (defined as that which the consumer takes home) but also the secondary or transit packaging and even the palletisation if this is relevant. Leaving the development of transit packaging until after all the primary components are finalised may lead to unnecessary and costly materials having to be used, and/or a late product launch.

In summary, the product and packaging development process is a multi-disciplined one, requiring information and decisions from almost all functions and operating processes within an organisation. There is a key role here for the proactive packaging designer, in both feeding ideas into the process and in developing innovative and cost-effective solutions. This means the designer should be involved from the outset, working as part of the development team together with the packaging technologist. The synergistic effect of co-operation between packaging designer and technologist, particularly at this early stage, cannot be overstated. It is hoped that the information and ideas presented in this work will encourage and inform both designers and technologists in this important role.

GENERAL REQUIREMENTS OF A CLOSURE

This section aims to provide the packaging designer with a checklist of the basic requirements which apply to all closures. It summarises the fundamental principles already discussed in detail and sets out some useful information on industry nomenclature.

The fundamental principles

The following demands apply to any closure:

1. It must effect a good seal to the container, from the packaging line through to the end of the useful life of the product.
2. It must be readily and safely removed from the container (and reapplied if required).
3. It must be readily handled on the filling line.
4. It must be chemically compatible with the product and with the container and comply with legal requirements such as food contact legislation.
5. It must be aesthetically acceptable as part of the overall product image.
6. It must meet the market requirements in terms of functionality and cost.

The extent to which some of these demands are met will depend on the dimensions (and the manufacturing tolerances) of the closure and the container to which it is applied. Fortunately, there are accepted industry, national and international standards for a wide range of closure types. In order to understand these, it is important to be familiar with the conventions for describing the important dimensions.

Understanding the terminology

First of all, the neck of a container is known as the finish. This is because in the early production of glass containers (which of course pre-dated plastic containers by some 3,400 years) the body of the bottle and the neck section were formed separately and the 'finish' or end of the process was to fuse the two together.

29

Types of thread finish

Thread finishes on containers can be of two main types: continuous thread, where the thread forms a continuous spiral around the neck, and interrupted thread, where the threads are formed in discrete sections:

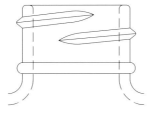

Continuous thread (CT) finish.   Interrupted thread or lug finish.

Continuous thread finishes are used on both narrow and wide-necked containers, while interrupted threads are used only on wide necks.

The continuous thread finish is available in a range of different standard heights, a general guideline being that the greater the diameter the shorter the neck height. The taller the finish, the greater the number of turns of thread which must be engaged, which has implications for capping machine operations as the more turns needed to apply the cap, the longer the time needed for this action.

For good seal efficiency, a minimum of one complete turn of thread engagement between the cap and the container is essential for continuous thread finishes. In this respect, one of the advantages of interrupted threads is that the threads can be satisfactorily engaged by a very short machine movement, often only one quarter of a turn, which means faster capping speeds. Another advantage is that the interrupted threads are less likely to become clogged with sticky product, and if they do so they are easy to clean – which is one reason why we see this type of finish used so frequently on jars of honey, preserves and pickles.

Caps and container neck finishes are both designated by size, e.g. 18mm, the number corresponding approximately with the 'T' dimension shown in the following diagram:

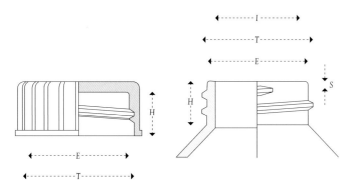

Cap dimension codes correspond to container finish dimension codes as follows:

For caps:
'T' is the diameter across the root of the threads.
'E' is the diameter across the tops of the threads.
'H' is the internal height (which must allow for a wad if one is to be used).

For containers:
'T' is the diameter across the tops of the threads.
'E' is the diameter across the root of the threads.
'H' is the height from the top of the finish to the shoulder.
'S' is the distance from the top of the finish to the start of the threads.
'I' is the internal diameter or bore.

There are agreed standards and tolerances for glass (BS 1918) and plastic (BS 5789) and container finishes and caps must be produced in accordance with these standards. The extremes of tolerance need to be considered. For example, the 'T' dimension of a cap must be controlled such that it can be readily applied without interference to a container which has a 'T' dimension at the maximum, but it should not be so loose that there is insufficient thread engagement when applied to a container with a 'T' dimension at the minimum.

Similarly, the internal height 'H' of the cap must not be so great that the cap hits the container shoulder before a seal is achieved on the top of the container (in which case the pack will leak), or so short that the cap sits too high above the container shoulder, which will be aesthetically unpleasing.

Thread profile

There are two common forms for the thread profile on containers:

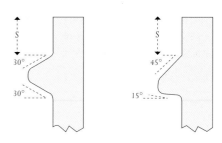

Symmetrical or rounded profile.   Modified buttress profile.

The symmetrical profile is the original shape designed for glass containers, and caps with this symmetrical profile were designed accordingly. With the development of plastic containers it became apparent that this profile is not really suitable for consistently making and holding a good seal. Thus the modified buttress thread has been developed as the preferred style for plastic container neck finishes.

Caps produced to the symmetrical profile will fit both the rounded contours on glass containers and the modified buttress thread on plastic containers, which is useful when using standard designs. However, if for some reason it is decided to develop a special cap thread profile for the modified buttress finish, this may not be suitable for application to a glass container.

Achieving a seal

The seal between a screw-threaded cap and container is usually achieved at one of two points. The simplest is the cap which seals to the top surface or 'land' of the container. This usually relies on a wad to take up the normal irregularities in the land surface (although this does not remove the need for control of the quality of this section during container production). A top seal can also be achieved using a soft, pliable flowed-in liner.

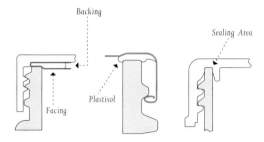

Wadded closure, top seal.
Flowed-in liner closure, top seal.
Linerless closure, bore seal.

More and more caps are now being developed where the seal is achieved on the inside of the container neck, as shown in the above diagram on the right. This approach places greater criticality on the control of the container neck dimensions (especially the 'I' dimension) but removes the need for a separate wad.

Torque

Torque is the turning effect of force upon an object and is an important consideration when applying a screw-threaded cap to a container. Caps are usually applied at pre-set torque levels and these should be controlled and checked on the packaging line, to avoid problems in use. Plastic closures always loosen (called 'backing-off') in the first 24 hours after application and this needs to be taken into account when setting the torque values. If the application torque is too low, the cap may not achieve an effective seal to the container. If it is too high, there is a risk of the cap overriding and thus damaging the threads, which will eventually lead to leakage; or of making the seal so tight that the pack cannot be opened using normal force, resulting in consumer frustration or even injury.

Icon key

 food & drink

 health & beauty

 household chemical

 industrial

 textiles

 tamper evidence

 child resistance

 measuring/dispensing

 award winner

 CD reference

31

32

MEASURING/ DISPENSING	FOOD & DRINK	HEALTH & BEAUTY	HOUSEHOLD CHEMICAL	INDUSTRIAL	TEXTILES	AWARD WINNER
				*	*	
					*	
					*	
*				*		
*			*	*		
*			*			*
*			*			
			*			
*			*			*
*	*					
*	*					
*		*				
*	*					
*	*	*				
*	*					
	*					
	*					
	*					
*		*				
	*					
	*					
		*				
*		*				*
		*				
	*					
	*					* ring-pull only
	*		*			
	*					
	*					
	*					
	*					
	*			*		*

34

MEASURING/ DISPENSING	FOOD & DRINK	HEALTH & BEAUTY	HOUSEHOLD CHEMICAL	INDUSTRIAL	TEXTILES	AWARD WINNER
	*					*
	*					
*			*	*		
	*					
	*					
*			*			*
*	*					*
	*					
	*					*
			*	*		
*	*		*			*
*	*			*		*
*			*			*
*		*				
*		*				*
		*				
	*	*				
		*				*
	*					
*		*				*
*	*					
*		*				
*		*				*
*			*			*
*		*				*
	*					
*		*				*
*		*				
*		*				
*		*				
*			*			
				*		

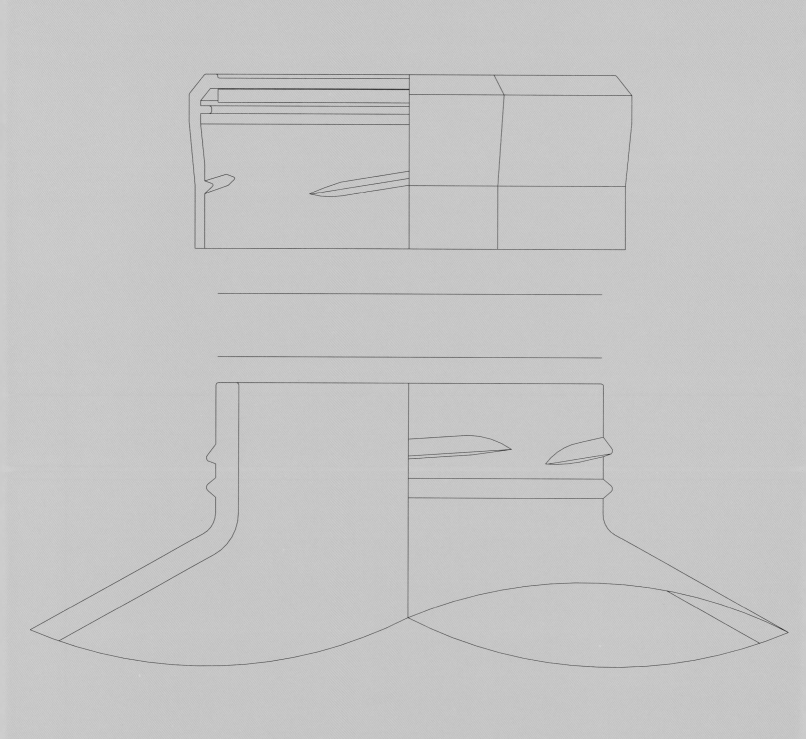

THE DESIGNS

RATIONALE FOR SELECTION OF CLOSURE
EXAMPLES

The closure examples included in this work have
been chosen to give the designer as wide a view as
possible of different packaging materials and
formats, and their uses in different markets. Some
have been selected for their traditional nature,
others to demonstrate some of the very latest ideas.
Importantly, it must be understood that inclusion
in this text does not imply recommendation and
exclusion does not imply criticism.

Similarly, while the authors have made every
effort to check the validity and accuracy of all the
information given throughout this work, no
responsibility can be accepted for any errors or
omissions. All information is provided on the
understanding that it must be verified by the user
before implementation. No responsibility is
accepted by the authors for any aspect of this
verification process.

The drawing shown with each design is intended to provide a visual appreciation of how
the closure works, to support the photograph and
brief description. All drawings are derived from
copyrighted material and must not be used for
commercial gain without the permission of the
design owner.

Anne and Henry Emblem
October 2000

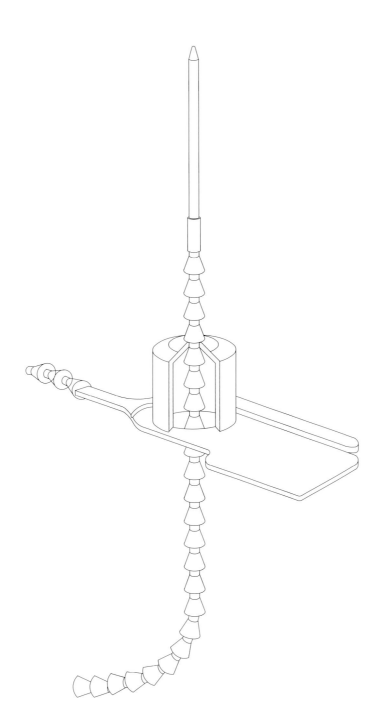

A tough plastic seal which is secured by feeding one end through a one-way locking mechanism. The seal can only be removed by cutting, showing visible evidence of tampering. Seals can be individually numbered for record purposes, giving increased security against pilferage or contamination. They are used to secure postal bags, tankers of food ingredients and any valuable or sensitive goods in transit.

PLASTIC SECURITY SEAL   39

ITW Fastex

This is a clever way of adding a display feature to a bagged product, for example clothing and accessories. The hook is easily fixed to the bag by means of two posts and clips. These spread the weight of the product across the bag and avoid excessive stress while the pack is hanging on display and being examined by customers. There is a label area for easy product identification.

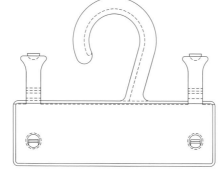

Marplan Ltd

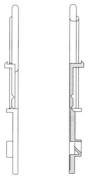

This is a very simple and effective way to make a conventional carton or case easy to display and carry. The handle simply slots into the top of the carton, and when folded back it gives a comfortable and easy shape for the hands to grip. The handle folds back during transit, which means it does not take up additional space.

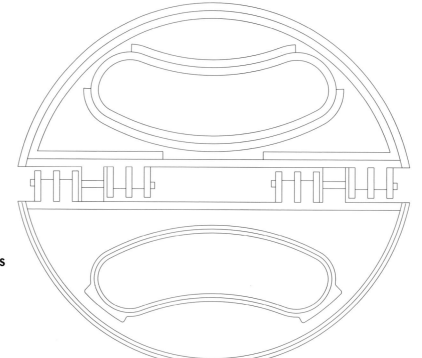

PLASTIC CARRY HANDLE FOR CARTONS

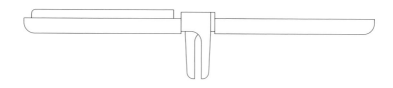

Morplan Ltd.

This pouring closure has a plastic central section and an outer metal ring for crimping onto tin plate containers. There is a tamper-evident ring on the plastic cap, which provides a convenient means of pulling up the extended pouring spout. Before opening, the closure sits flush in the top of the container.

PRESS-FIT AND CRIMPED
EXTENDABLE POURING CLOSURE   45

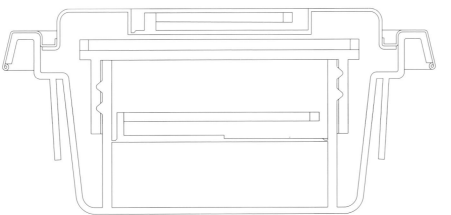

Designed to be screwed on to the neck of a plastic container, this closure has a tamper-evident feature to denote first opening. The spout is pulled up to provide a handy extended pouring device, allowing the product to be dispensed into difficult-to-reach areas, making the system especially useful for products such as motor oils. An air-vented flow control system optimises pouring, giving smooth and non-spill dispensing.

46 **SCREW-THREADED EXTENDABLE POURING CLOSURE**

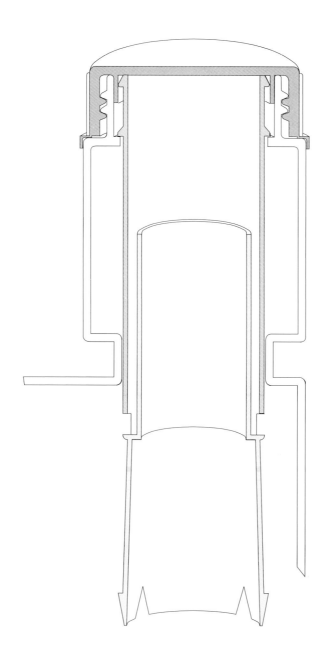

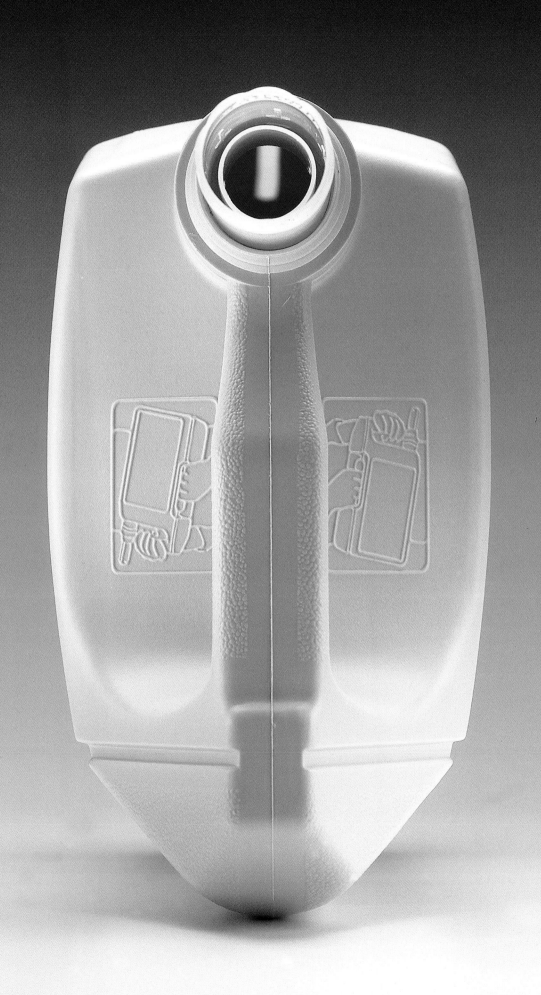

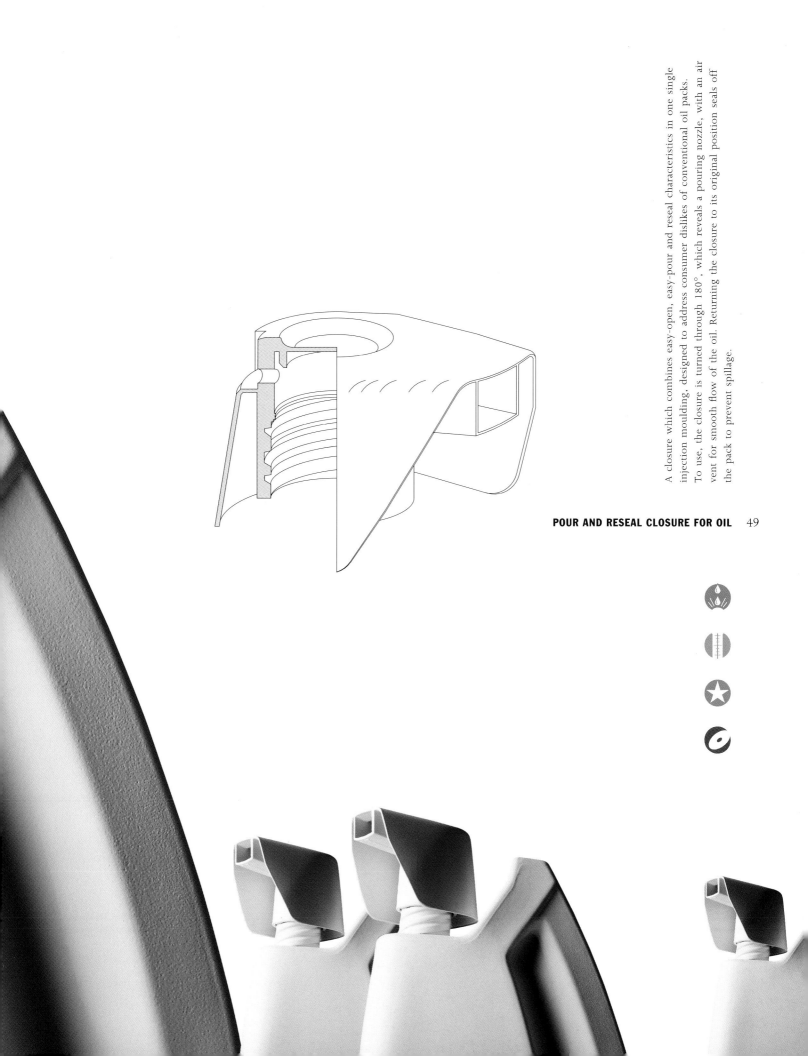

A closure which combines easy-open, easy-pour and reseal characteristics in one single injection moulding, designed to address consumer dislikes of conventional oil packs.

To use, the closure is turned through 180°, which reveals a pouring nozzle, with an air vent for smooth flow of the oil. Returning the closure to its original position seals off the pack to prevent spillage.

POUR AND RESEAL CLOSURE FOR OIL   49

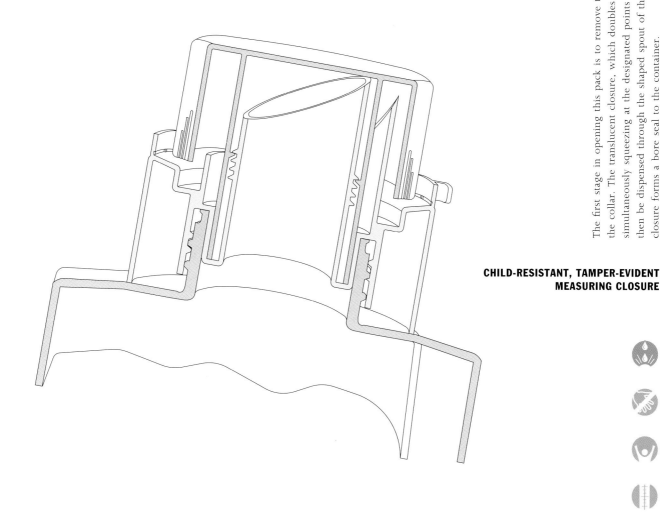

The first stage in opening this pack is to remove the tamper-evident plastic band from the collar. The translucent closure, which doubles up as a measure, is then removed by simultaneously squeezing at the designated points and turning. The liquid product can then be dispensed through the shaped spout of the collar, into the measure. The closure forms a bore seal to the container.

CHILD-RESISTANT, TAMPER-EVIDENT MEASURING CLOSURE

51

This fabric paint pack consists of bottle, nozzle fitment and overcap. The nozzle is an interference fit to the bottle via mating beads. This fit is critical as the nozzle must be pushed in during filling without crushing the bottle, stay in place under application pressure, and be removable for unclogging if the overcap is left off. The overcap seals the nozzle orifice, provides easy display and allows the pack to stand inverted to keep the paint in contact with the applicator tip.

PAINT APPLICATOR

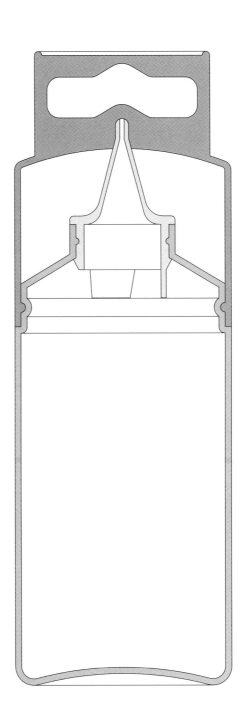

Rexam Closures and Containers, USA

Dragon Plastics Ltd.

The Savalok® closure system opens on the 'squeeze and turn' principle. To open, the cap is squeezed at the two designated pads, which disengages it from the two locking features on the container neck and allows it to be unscrewed. This requirement for two simultaneous actions is a key feature of child-resistant closures.

'SQUEEZE AND TURN' CHILD-RESISTANT CLOSURE   55

This Japanese pack contains a pre-wash laundry liquid which is dispensed in a mousse format through a two-part closure with a dip tube. The top part of the closure is turned to open, giving an audible 'click' when the correct location is reached. Squeezing the plastic bottle then dispenses the product in a jet of foam through the angled nozzle, allowing it to be directed onto the soiled laundry. Turning the top part of the closure closes off the nozzle, again giving audible confirmation.

56   **SPREADING CLOSURE FOR MOUSSE**

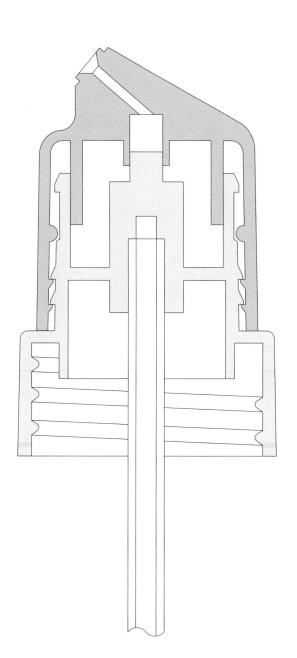

Lion Corporation, Japan
PackTrack™ CPS International Ltd.

This is one of three Japanese packs chosen for inclusion in this work due to their innovative features. The closure has a very smooth and easy-to-operate push button which causes a hinged lid to flip up, revealing dispensing holes for sprinkling the product. The smooth action allows one-handed operation while cooking and the hinged section is easily pushed to close. Small nibs moulded into the underside of the lid seal up the dispensing holes, preventing product deterioration and spillage.

58 **FLIP-TOP CLOSURE FOR POWDER DISPENSING**

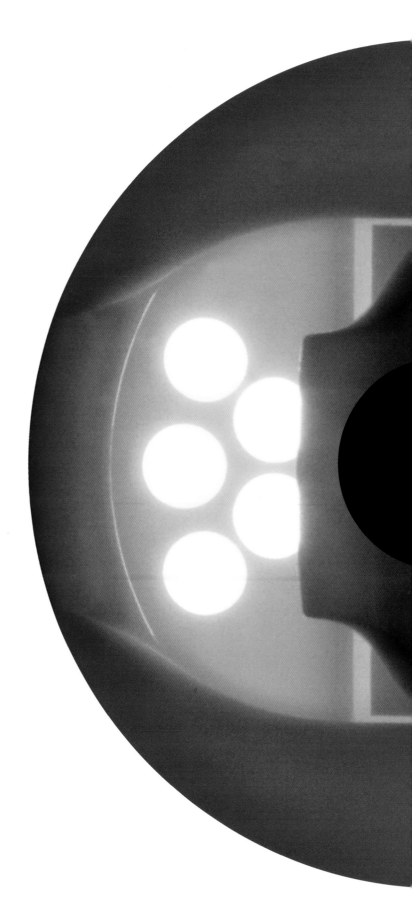

House Foods Corporation, Japan
PackTrack™ CPS International Ltd.

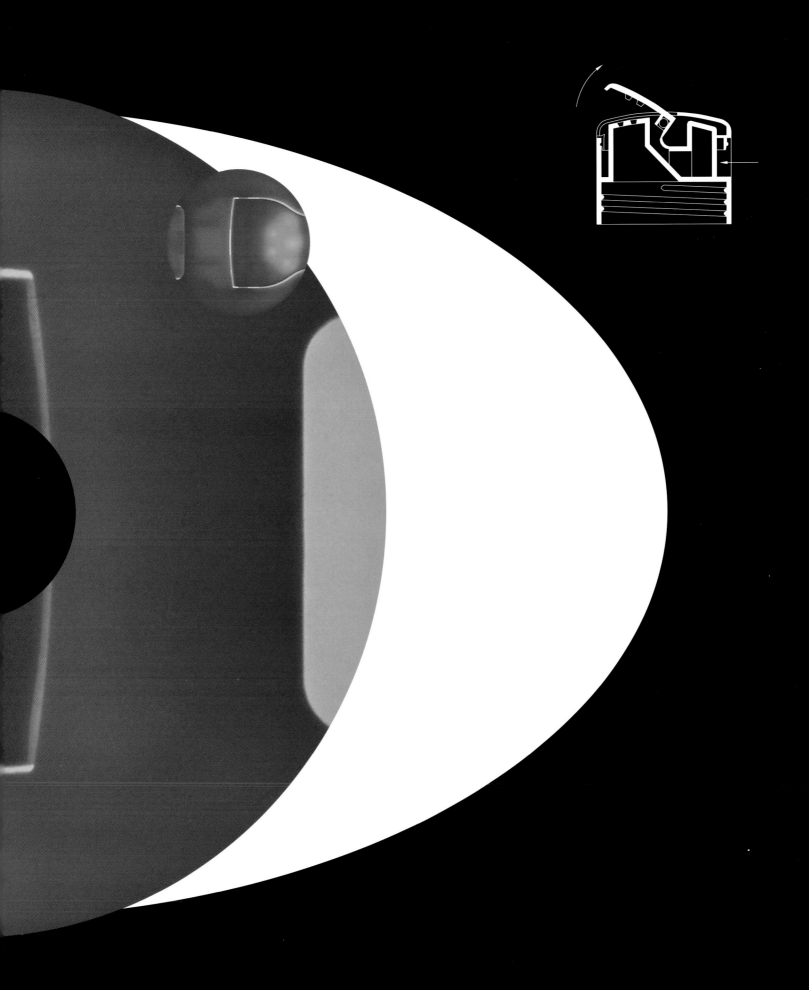

A push-in plastic closure which seals the pack and provides a useful means of dispensing the powdered product through the special sprinkler opening. The orifice can be sealed off after use by a rotating disc, thus maintaining product freshness. The sprinkler is recessed and thus does not take up additional space during transport and storage, it also allows the pack to be stacked neatly.

TURN-TOP POWDER SPRINKLER CLOSURE   61

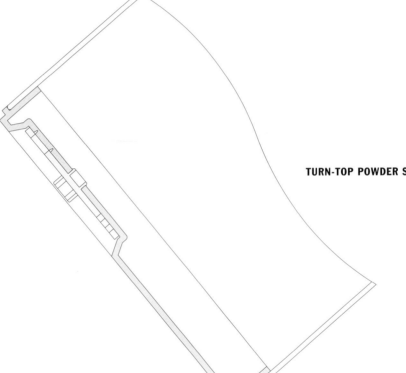

Sonoco Consumer Products Ltd

Dragon Plastics Ltd.

A screw-threaded, bore seal cap with hinged lid, moulded in one piece. The top is easily flipped up and the product dispensed through the nozzle in a one-handed operation. The 'live hinge' construction is made possible by the use of polypropylene, which has excellent resistance to repeated flexing.

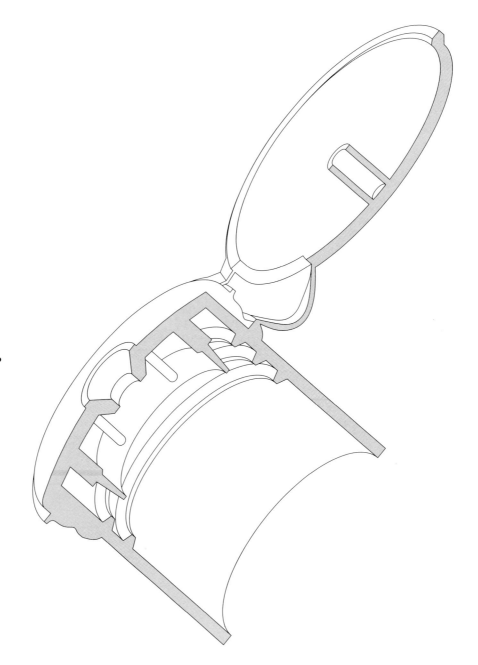

A three-piece closure system offering two tamper-evident features for product integrity, a bore seal to prevent leakage, and a pull-up spout for easy operation 'on the run'. The first tamper-evident system is a perforated band around the base of the screw-thread part, which breaks away on any attempt to unscrew the whole closure. The second is the seal of the clear overcap which has to be removed to pull up the spout and gain access to the product.

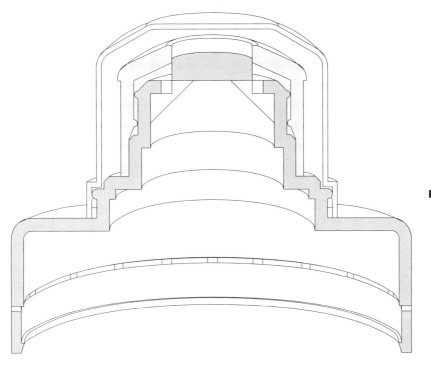

PULL-UP CAP FOR SPORTS DRINKS 65

Betacap UK Ltd

The SimpliSqueeze® closure concept uses a flexible membrane which allows product flow only when pressure is applied to the pack by squeezing. This is useful for still beverages such as sports drinks and mineral water: the pack can be opened with one hand and the drink consumed 'on the go' without spilling. The concept also allows the pack to be stored upside down, a feature which is utilised in hanging packs for shower gels etc.

Seaquist-Löffler Ltd

66 **VALVE DISPENSING CAP**

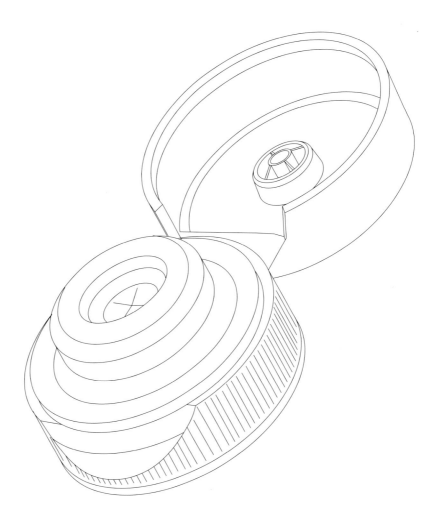

A two-piece closure for a press-fit to standard-neck-finish glass or plastic bottles. The outer cap is removed by unscrewing, to reveal a tear-out tamper-evident membrane to be removed at first use. The product flow is controlled by a built-in regulator and a Droples® pouring feature of flexible plastic 'fins'. The tamper-evident membrane provides an enhanced barrier to gases for extended shelf life of the product.

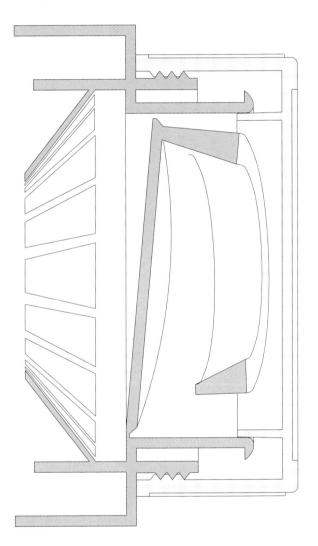

Pericap UK Ltd

DRIP-FREE DISPENSING CLOSURE 69

This cap is fitted with a foamed plastic wad, wax-bonded to aluminium foil which has a heat-sealable coating. The cap is screwed down and the pack passes under an induction coil, generating a magnetic field which induces heating currents in the aluminium. The heat softens the heat-sealable coating, sealing the foil to the top of the jar. On opening, the wax layer shears and the plastic wad is retained inside the cap with the foil providing a tamper-evident, moisture- and flavour-barrier diaphragm.

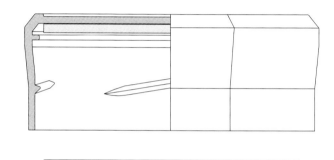

Nestlé UK Ltd

70 **INDUCTION-SEALED DIAPHRAGM
CLOSURE SYSTEM**

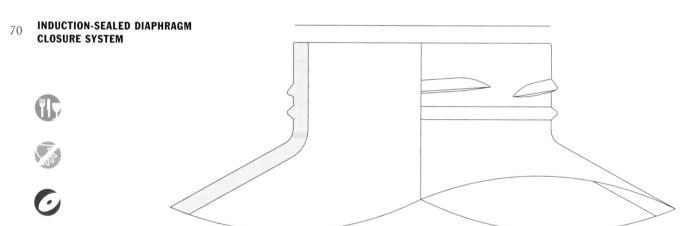

Metal Packaging Manufacturers Association Ltd.

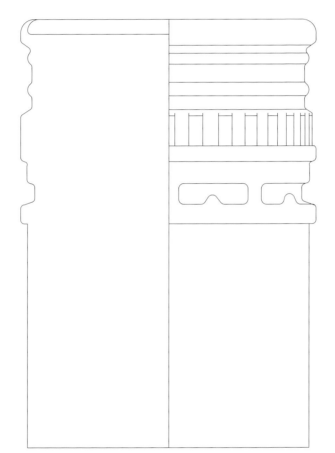

The ROPP cap is made as a printed aluminium shell with a perforated lower ring and is fitted with a wad to form the sealing surface to the glass bottle. It is rolled onto the bottle using a special tool which shapes the soft metal into the threads of the bottle, and folds it around a neck ring on the bottle. On opening, the perforated section drops down under the neck ring and becomes trapped on the bottle, thus giving visible evidence of previous opening.

ROLL-ON PILFER-PROOF ALUMINIUM CAP · 73

Non-refillable fitments offer anti-counterfeiting measures, especially in the high-value spirits sector. To be effective they have to balance two requirements: they must be difficult to bypass, yet not impede the flow of the liquid when being poured from the bottle. Their construction is such that attempts to remove them from the bottle result in breakage, leaving parts of the fitment trapped in the bottle neck.

74 **NON-REFILLABLE CLOSURE SYSTEM**

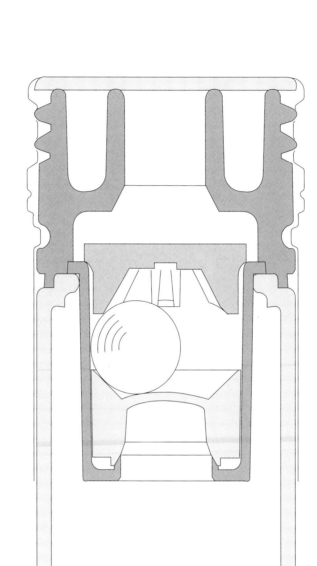

United Closures and Plastics

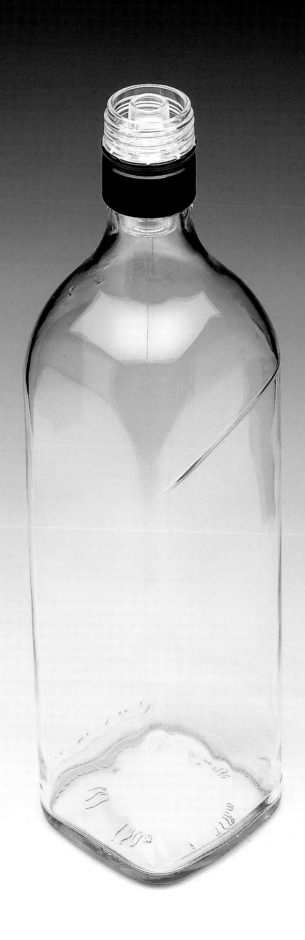

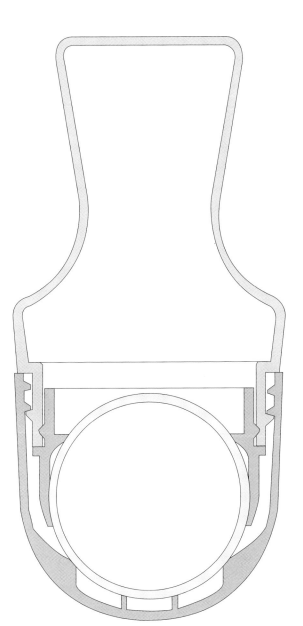

A simple way of applying a viscous liquid is by using this roll-on closure system. The plastic ball fits into the top of the container and acts as the applicator. The fit of the ball in the bottle housing must be loose enough to rotate, but not allow product leakage. The pack is sealed by the overcap pressing down on the ball so that it sits on the sealing ring around the inside of the bottle neck.

ROLL-ON BALL CLOSURE SYSTEM 77

The first tamper-evident feature on this pack is the perforated plastic band which breaks away on first opening and falls down the neck of the jar. The second is the central button which makes an audible 'pop' on first opening and thereafter can be depressed to give the distinctive clicking sound. The seal is achieved when the flowed-in sealing compound in the metal shell softens during the process of heat-sterilising the product, conforming closely to the threads on the neck of the jar.

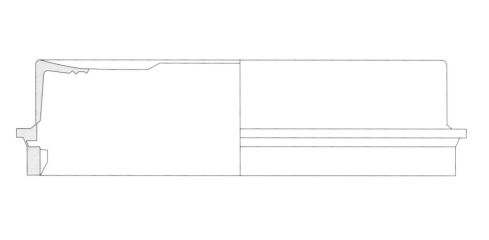

78 **DOUBLE TAMPER-EVIDENT TWIST-OFF METAL CAP**

This type of closure is made from tin plate and has small lugs instead of screw-threads, and a flowed-in sealing compound. It is used on hot-filled products. On cooling, the metal is pulled down into a concave shape, due to the partial vacuum inside the pack. This is released on opening and the metal springs up with an audible 'pop'. Replacing the closure leaves the clearly visible domed or convex top which can be pressed down, giving a distinctive sound.

The secure sealing of vials is of vital importance in the pharmaceutical and healthcare sector, while at the same time the filled and sealed pack should be easy and safe to use. In this system a glass vial is fitted with a rubber stopper, which is secured in place by a crimped aluminium shell. A plastic flip-up cap allows the closure to be easily removed, at the same time providing tamper evidence. The system withstands autoclaving if required.

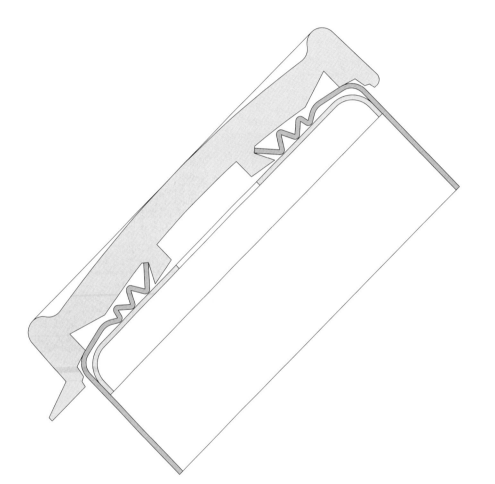

West Pharmaceutical Services

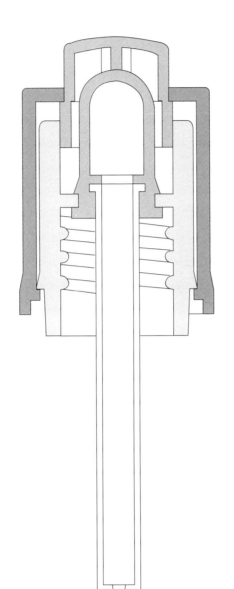

'Dropalok' is a screw-threaded closure system which fits onto a standard-neck-finish container and can be used for products such as aromatherapy oils and pharmaceutical preparations. Once applied, the closure can only be opened by both pushing down and turning the main section, which gives the child-resistant feature. The central section is an integral dispenser which delivers a measured drop when depressed.

CHILD-RESISTANT DROPPER DISPENSER 83

An ampoule made from high-quality, neutral tubular glass, for pharmaceutical injection products and cosmetic items. After filling, the open end of the ampoule is sealed by fusing the glass at high temperature, thereby producing a hermetic pack and giving a long shelf life to the product. If required, the filled ampoules can be sterilised. To open, the neck of the ampoule is snapped cleanly along the special break ring.

84 **GLASS AMPOULE WITH HEAT-FUSED CLOSURE**

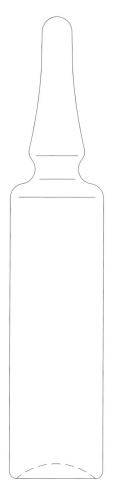

Adelphi (Tubes) Ltd.

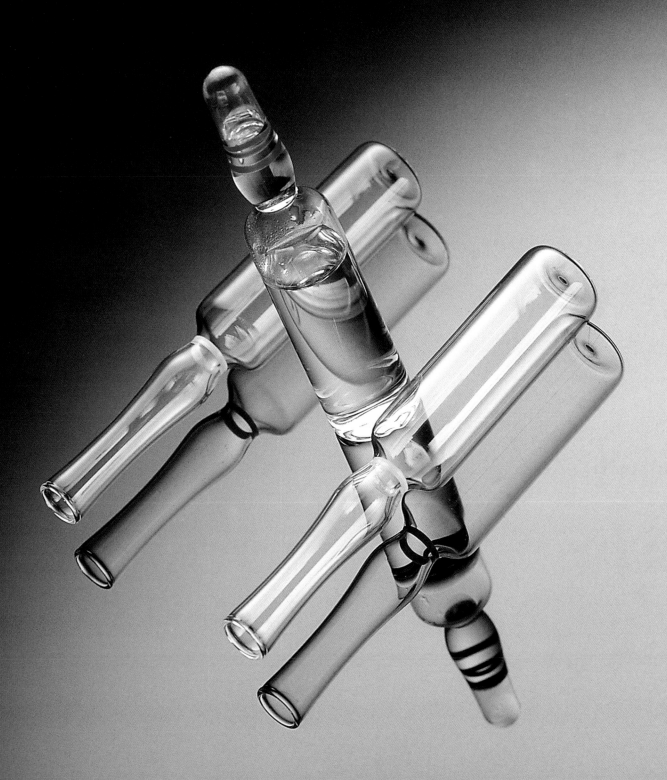

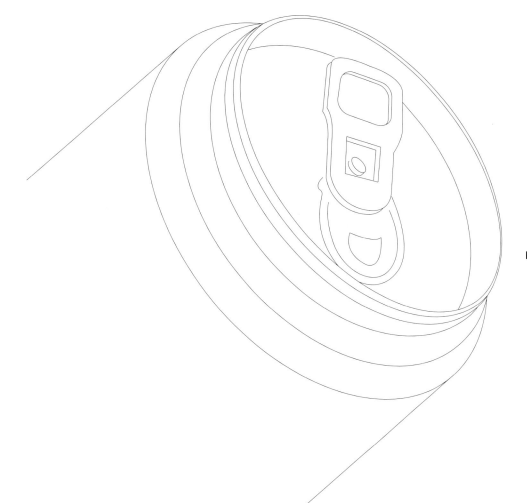

A ring-pull end which is seamed on to the filled can using a double seam process, to give a hermetic pack. The filled can withstands the pasteurisation process usually required for beer products. To open, the ring is pulled to give a small aperture through which the product can be poured or drunk directly. The ring-pull remains attached to the can, thus helping to reduce litter.

RETAINED RING-PULL DRINKS CAN   87

Metal Packaging Manufacturers Association Ltd.

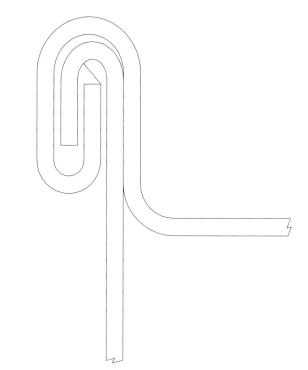

Food cans are traditionally closed using plain metal ends and opened with a tin-opener. In response to consumer demand, many are now fitted with full-aperture ring-pull ends, which can be removed by hand without special tools, as the EOLE II sample shown here. Whichever type is used, it is sealed onto the can after filling using a double seam process to ensure that a hermetic pack is obtained. Along with the sterilisation process, this gives canned food its long shelf life.

Metal Packaging Manufacturers Association Ltd.
Carnaud Metalbox Food UK

CONVENTIONAL AND RING-PULL END METAL CANS

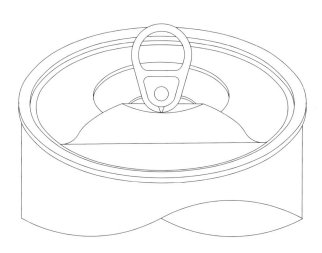

A pressed metal lid which forms a friction seal when pushed into a metal ring seamed onto the top of the container. The lid has to be levered up at the edge to open it. This type of pack is commonly used for paint, as the opening is easy to keep clean. It is also used for foods such as custard powder and catering packs of coffee, when a diaphragm seal is usually added, for moisture barrier and tamper resistance or evidence.

90   **METAL OR COMPOSITE CAN WITH LEVER LID**

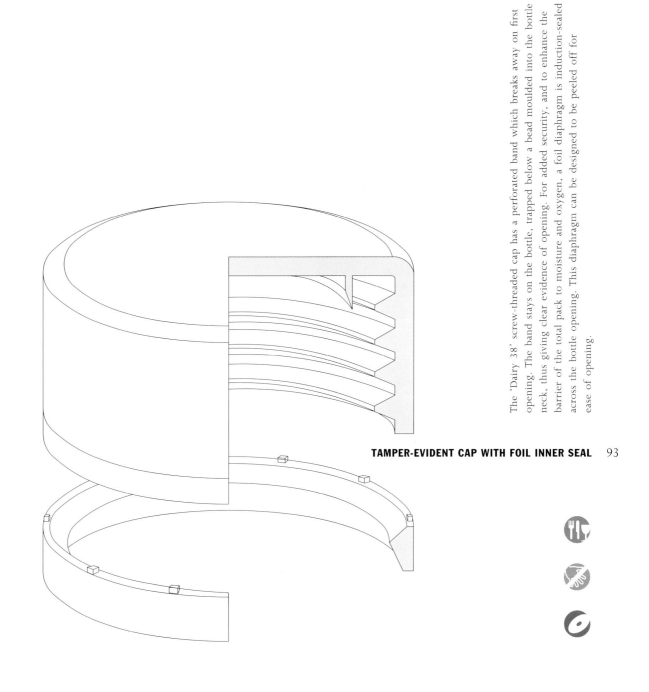

The 'Dairy 38' screw-threaded cap has a perforated band which breaks away on first opening. The band stays on the bottle, trapped below a bead moulded into the bottle neck, thus giving clear evidence of opening. For added security, and to enhance the barrier of the total pack to moisture and oxygen, a foil diaphragm is induction-sealed across the bottle opening. This diaphragm can be designed to be peeled off for ease of opening.

TAMPER-EVIDENT CAP WITH FOIL INNER SEAL   93

The closure is a hinged polypropylene moulding secured to the top rim of the container, with a tamper-evident diaphragm sealed across the opening. The diaphragm also maintains a barrier to moisture and gases, to preserve the product until its first opening. The hinged lid can be opened with one hand and snaps firmly in place with an audible and reassuring 'click' to close the container. The wide aperture allows for easy product access.

94 RECLOSABLE PACK WITH HINGED LID

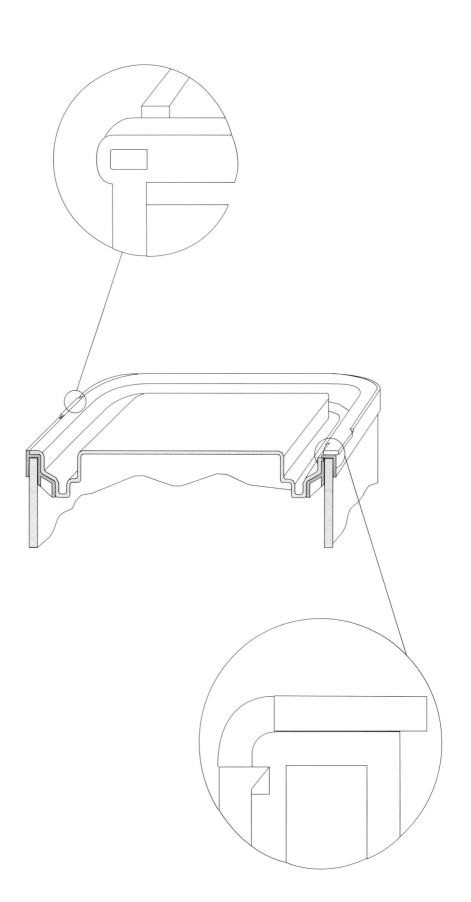

Akerlund and Rausing Ltd.

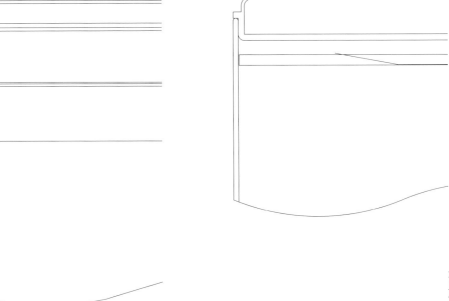

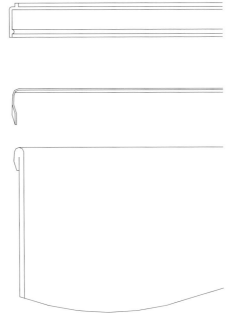

There is a wide range of closure systems available of this type. The diaphragm seals provide tamper evidence and barrier properties to moisture, oxygen and volatiles, and can be designed to be peeled off or punctured, to meet customer requirements. The snap-on or push-in plastic lids provide the reseal feature once the pack has been opened, maintaining freshness to the end of the product's shelf life.

Sonoco Consumer Products Ltd.

96 **DIAPHRAGM SEALS WITH RESEALABLE LIDS**

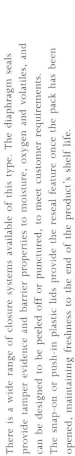

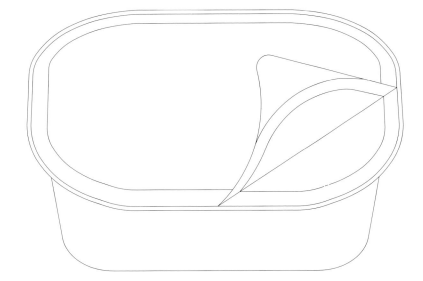

Alcan Deutschland GmbH

A drawn aluminium container with a sealed lid which can be peeled off without the need for any special opening tools. The product is a single serving and resealing is thus not required.

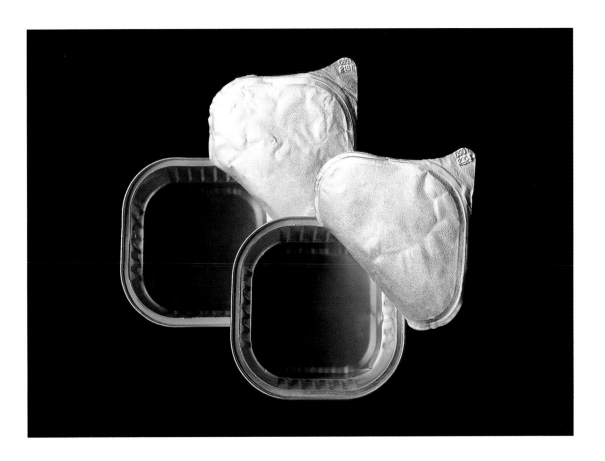

The sack is opened the first time by pulling a ripper tape sewn across the top. It can then be reclosed by means of the patented zipper feature, which is robust enough to survive the life of the product. The system is easily added to sacks for use on existing filling equipment. The sacks can be paper or plastics and can be carried via the attached plastic handle.

Zip Pack Packaging Technology Ltd.

pull here

tirar aquí

hier ziehen

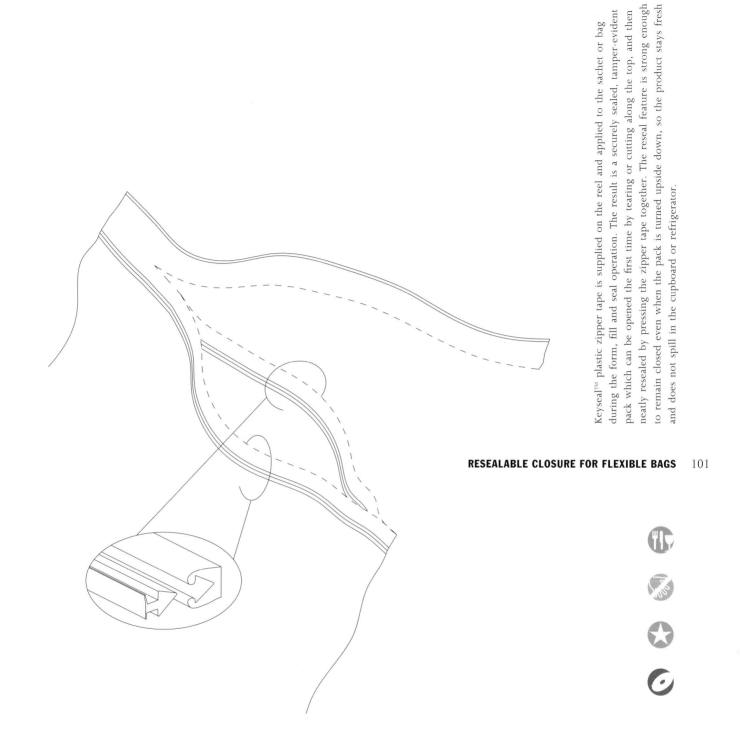

Keyseal™ plastic zipper tape is supplied on the reel and applied to the sachet or bag during the form, fill and seal operation. The result is a securely sealed, tamper-evident pack which can be opened the first time by tearing or cutting along the top, and then neatly resealed by pressing the zipper tape together. The reseal feature is strong enough to remain closed even when the pack is turned upside down, so the product stays fresh and does not spill in the cupboard or refrigerator.

RESEALABLE CLOSURE FOR FLEXIBLE BAGS 101

This is the Swedish-designed Reseal-it™ system used for flexible packs of cheese, cooked meats etc. The pack addresses many of the consumer complaints about flexible packs for this type of product, which are often difficult to open without using scissors or a knife. Reseal-it™ offers both easy opening via a peelable seal, and an effective way of closing the pack by simply resealing the special label. Unused product can thus be conveniently and safely stored in the refrigerator.

102 **PEELABLE AND RESEALABLE LABEL CLOSURE SYSTEM**

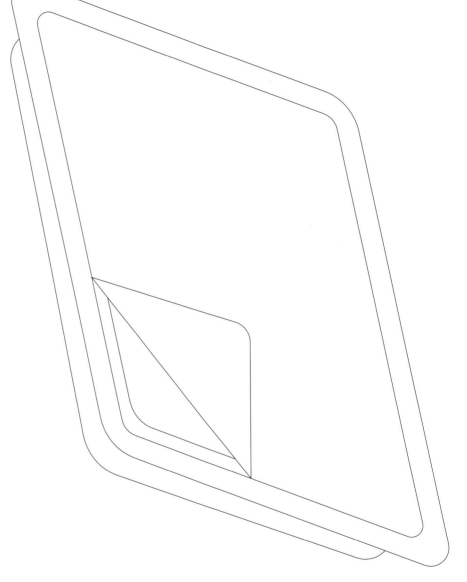

Reseal-it Sweden AB

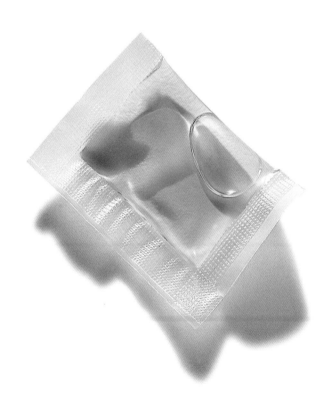

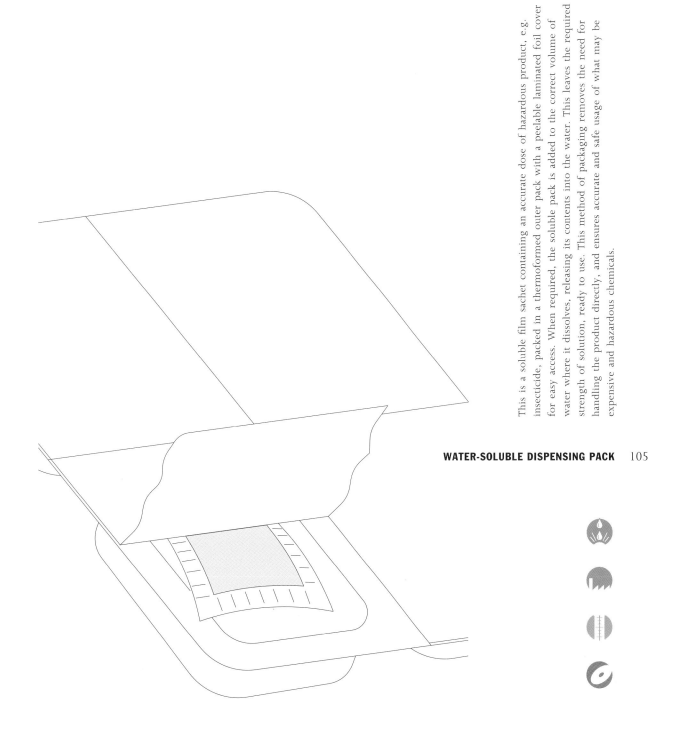

This is a soluble film sachet containing an accurate dose of hazardous product, e.g. insecticide, packed in a thermoformed outer pack with a peelable laminated foil cover for easy access. When required, the soluble pack is added to the correct volume of water where it dissolves, releasing its contents into the water. This leaves the required strength of solution, ready to use. This method of packaging removes the need for handling the product directly, and ensures accurate and safe usage of what may be expensive and hazardous chemicals.

WATER-SOLUBLE DISPENSING PACK   105

This lid is locked in place and can only be removed by breaking off a small tab of plastic on the rim. This allows the lid to be easily lifted up and removed, but leaves clear evidence of opening. For added security and product preservation, a film or foil diaphragm, which has to be peeled off, can be heat-sealed to the top of the container.

Peerless Plastics Packaging

106   **ROUND PLASTIC TUB WITH LOCKING LID**

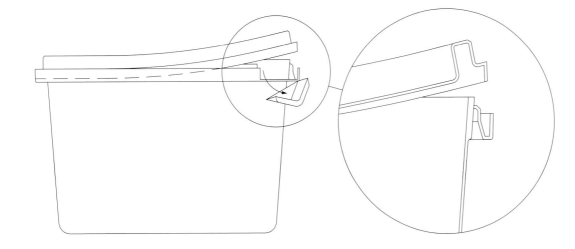

Peerless Plastics Packaging

A patented tamper-evident system known as Evilock™. The lid is snapped into place on the container and can only be removed by breaking a tab on one corner. This means the lid will not come off accidentally during transit, and when it has been removed there is clear evidence that the pack has been opened because the corner tab is missing.

This innovative paint pack has several convenience features: a wide aperture for large paintbrushes, a pouring feature via each corner, a square edge against which to wipe the brush, a retractable carry handle and an easy-open closure. It also survives dropping without spillage. When the pack is dropped the paint hits the central panel in the soft closure, causing it to flex upwards and outwards, which actually tightens the closure grip on the corners of the container and thus the pack remains intact.

108   **DROP-PROOF PACK WITH CARRY HANDLE**

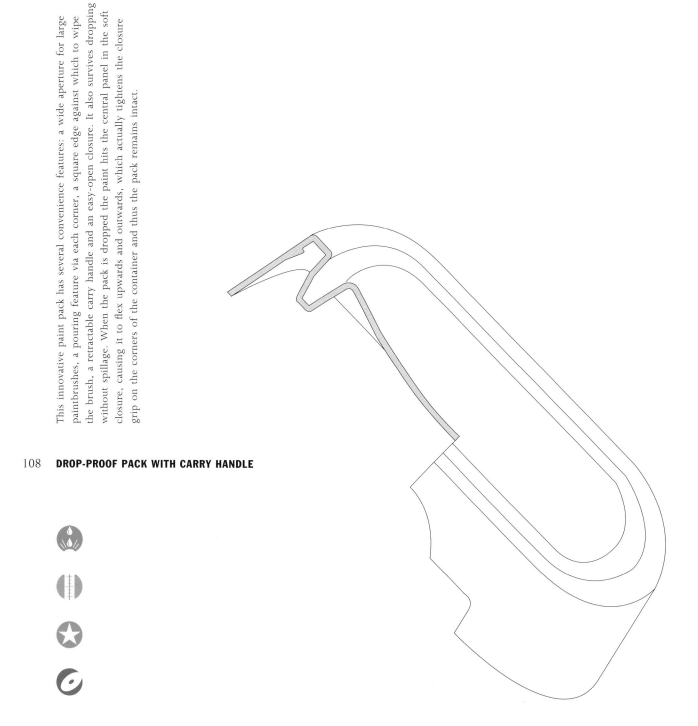

This is a novel and eye-catching two-piece pack featuring a unique 'Pop-Lock' opening device. A small lid flips up, making a 'popping' sound when pressure is applied at the point indicated in the moulded structure. The contents can then be dispensed in a one-handed operation, thus making the pack very easy to use when 'on the move'.

110 **POP-UP DISPENSER**

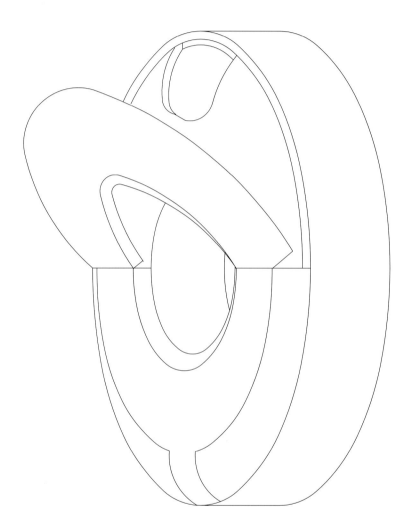

RPC Containers Ltd., Market Rasen

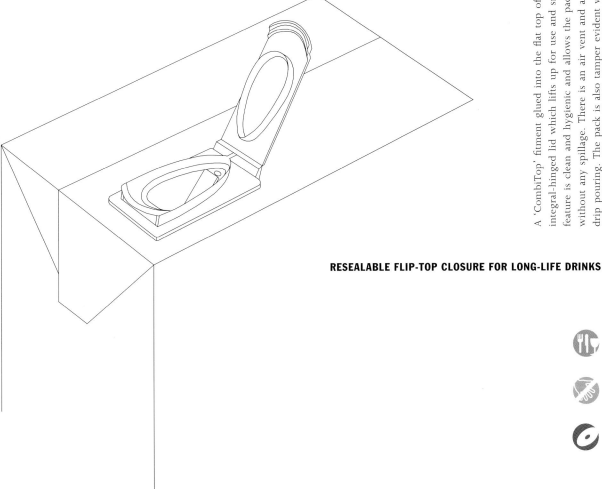

A 'CombiTop' fitment glued into the flat top of a drinks carton. The fitment has an integral-hinged lid which lifts up for use and snaps closed to reseal the pack. The reseal feature is clean and hygienic and allows the pack to be shaken or tipped once opened, without any spillage. There is an air vent and a specially designed lip for smooth, non-drip pouring. The pack is also tamper evident via a press-through plastic tab.

RESEALABLE FLIP-TOP CLOSURE FOR LONG-LIFE DRINKS   113

TetraTop® Mini GrandTab closure for drinks cartons is developed especially for the snack market. The plastic tab opens easily to reveal a large, hygienic aperture for drinking from directly. The square shape is comfortable and easy to hold. The plastic top section is moulded onto the paperboard body (produced from reels of printed material) immediately prior to filling, which means there is no requirement for large storage space for empty containers.

EASY-OPEN INTEGRAL PLASTIC-TOPPED CARTON

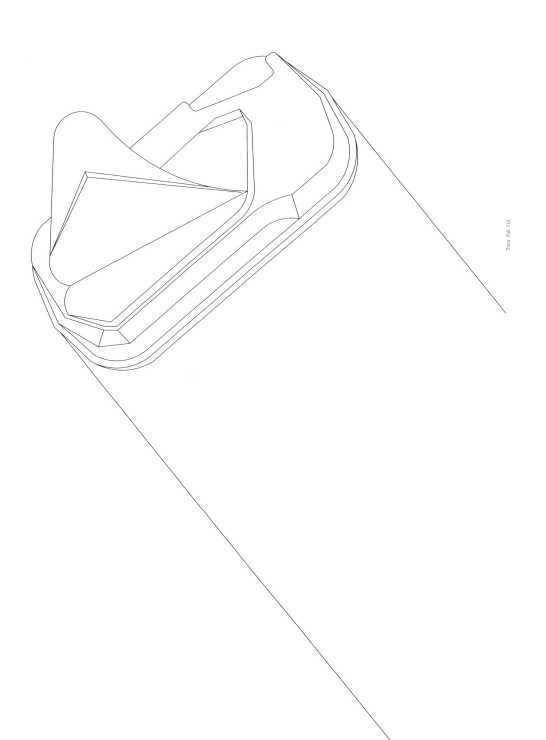

Tetra Pak Ltd.

Rippatape© is a heavy-duty tape for easy opening of sealed corrugated and fibreboard cases and cartons without using knives, thus preventing product damage. The end of the tape is secured in a board tab, which provides a means of gripping the tape to tear open the pack cleanly and efficiently. Applications include detergent cartons, document envelopes and outer cases of goods for display in supermarkets, where safe, fast and clean opening is important. Supastrip© (not shown here) is a light-weight tape for the easy opening of consumer packaging and is suitable for roll-wrap, overwrap, flow wrap, vertical form-fill seal and shrink sleeve applications. Available in widths from 1.6–12mm, the tape is self-adhesive and so easy to apply. Supastrip© can be printed in up to eight colours for high impact shelf visibility, branding or promotional themes. There are many applications including confectionery, biscuits, pharmaceuticals, stationery, tea and audio or video products.

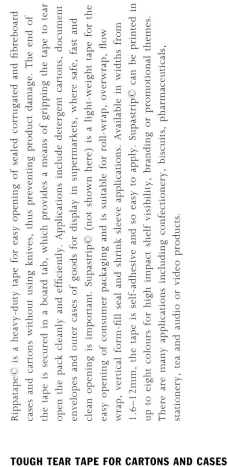

PP Poynx

TOUGH TEAR TAPE FOR CARTONS AND CASES

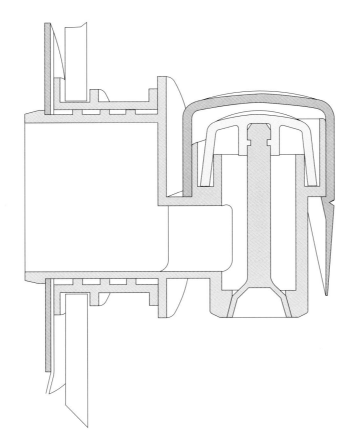

A self-closing, one-hand operated Press Tap for use with flexible Bag-in-Box® packaging. There is a tamper-evident cover which has to be peeled off before allowing access to the tap. The closure is hygienic and non-drip, and allows excellent control over the flow of liquid. Taps are available in different materials and with different fitments, making them suitable for a wide range of applications.

PLASTIC PRESS-ACTION DISPENSING TAP   117

A robust, screw-action, non-drip plastic tap for industrial and domestic applications. Taps are available for food use (e.g. home-brewed beer), chemicals and petroleum-based products, and with a variety of fittings, making the concept very flexible across a range of different end-uses. They can be fitted to both plastic and steel drums.

118 **PLASTIC SCREW-ACTION DISPENSING TAP**

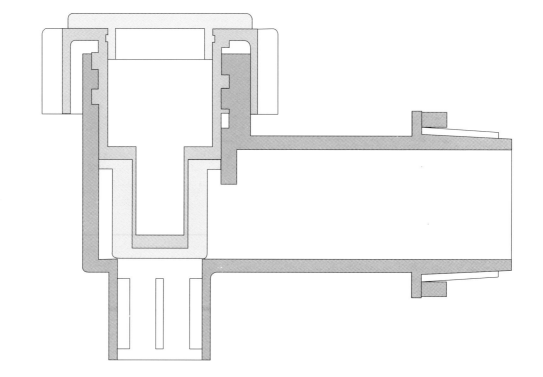

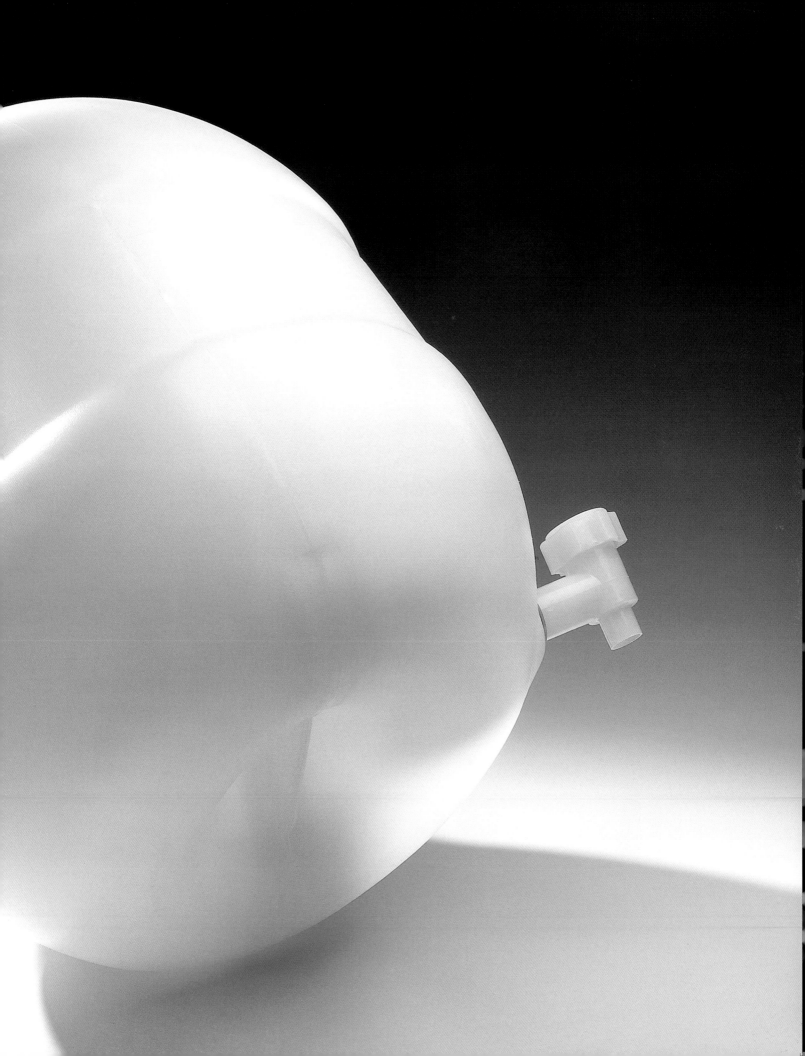

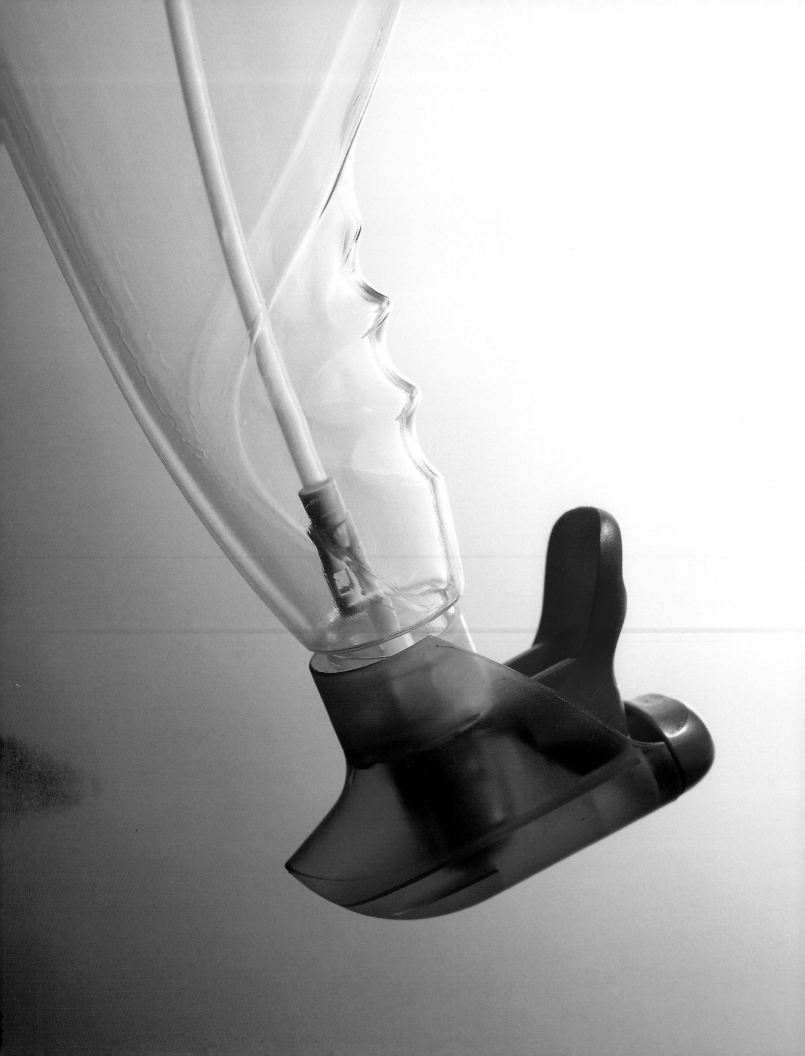

P.I. Design International. Developed and manufactured by Guala Dispensing S.R.L.

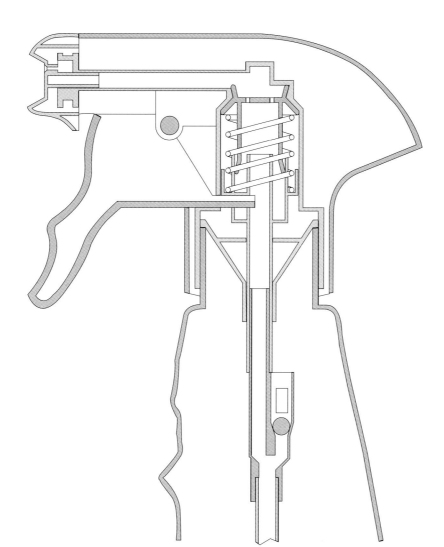

This is a cleverly designed trigger spray which can be used at any angle, even upside down, thus allowing the product to be directly applied to difficult-to-reach places. This makes it especially useful for the safe and convenient handling of domestic cleaning fluids.

MULTI-USE TRIGGER SPRAY 121

Techpack UK Ltd.

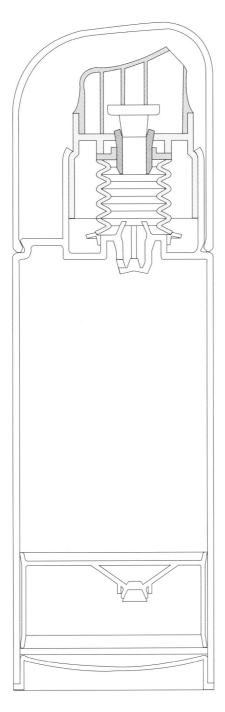

An all-plastic airless pump dispenser system housed in a cylindrical or oval case for good on-shelf appearance and stability in use. The pump is easy to depress and dispenses an accurate amount of product and the airless system makes sure the product is kept clean and hygienic until required. The pack is easy to fill on standard filling equipment and uses no propellants.

OVAL AIRLESS PUMP DISPENSER   123

This patented clikpak™ for homeopathic remedies dispenses a single pillule in response to one audible 'click' on the button. The pillule falls into the dosage cap, and can then be tipped into the mouth without being touched by hand. This is important as homeopathic remedies can lose their effectiveness when touched. A stepped funnel ensures that the pillules do not jam in the mechanism and the translucent pack means that the number remaining is easily seen.

Maddison Product Design and A. Nelson & Co. Ltd.

124 **DISPENSER FOR SINGLE PILL DOSAGE**

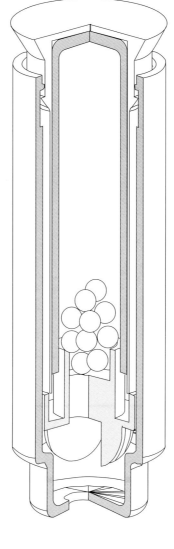

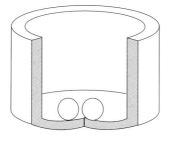

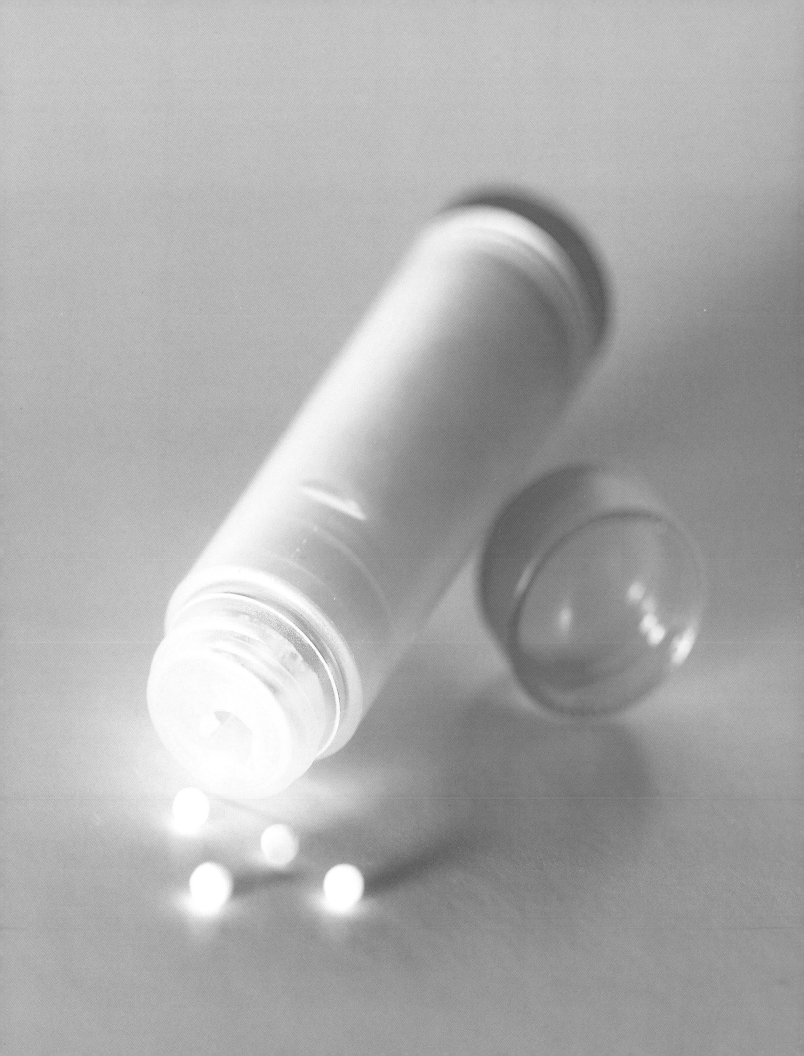

The 'Kidlok' is a two-part wadded closure, consisting of an inner threaded cap and an outer shell. It can only be removed by two simultaneous actions of pushing down and turning. The pressing action engages the outer shell with the inner cap, which then allows it to be unscrewed. Attempts to unscrew the closure without pressing down result in a series of audible clicking sounds.

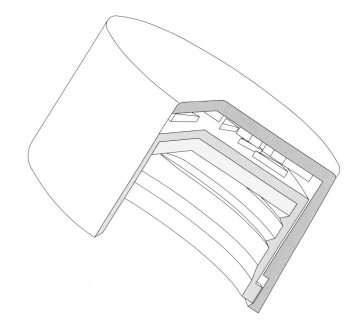

Dragon Plastics Ltd.

126 **'PUSH DOWN AND TURN' CHILD-RESISTANT CLOSURE**

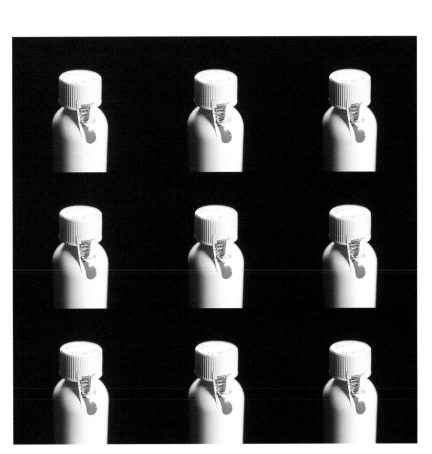

This is a patented security device called PULSLINE®. The label is attached to the container using an ultra-strong adhesive which means that attempts at removal effectively destroy it, and the use of a banknote thread in the label construction make it extremely difficult to replicate and replace. The thread can be made visible to the naked eye, or only visible under UV light, or only visible after peeling off a protective top-film, and it provides a simple and effective way of checking authenticity of the product and pack.

Inteac Ltd.

TAMPER-EVIDENT, ANTI-COUNTERFEIT LABEL   127

The 'Tampalok' combines tamper evidence and child resistance. It is a two-part closure, consisting of an inner threaded cap and an outer shell. It can only be removed by the two simultaneous actions of pressing down and turning, which provides the child resistance. The pressing action engages the outer shell with the inner cap, which then allows it to be unscrewed. The inner part has a perforated skirt which breaks on first opening and falls away, showing clear evidence of tampering.

Dragon Plastics Ltd.

128 **CHILD-RESISTANT SCREW CLOSURE WITH TAMPER EVIDENCE**

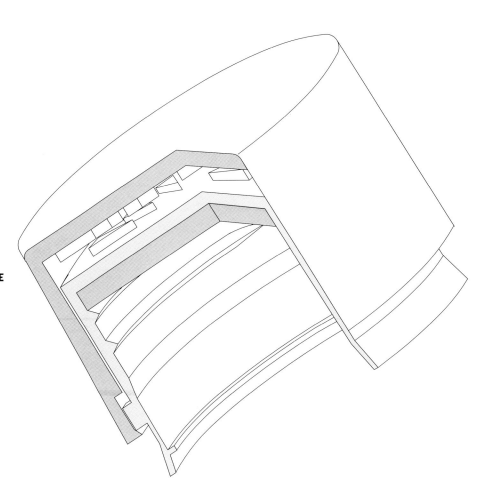

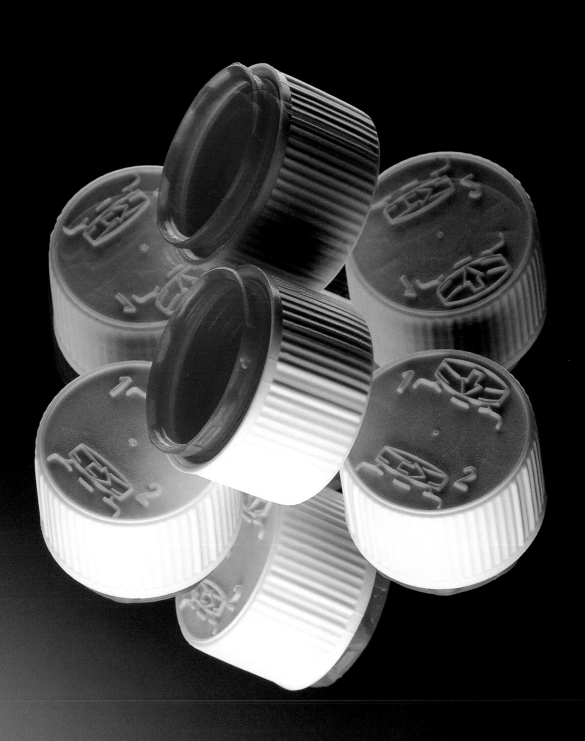

A 'SpoutCap' fitment sealed by ultrasonic welding into the angled top of a drinks carton. The fitment has a tamper-evident plastic ring which has to be lifted up before the cap can be unscrewed. The spout allows the drink to be poured out smoothly and the cap is easily screwed in place to reseal the pack for short-term refrigerated storage.

RESEALABLE SCREW-TOP CLOSURE FOR FRESH DRINKS

131

Rexam Combiblow Ltd.

Collapsible tubes are widely used for toothpaste, pharmaceuticals and cosmetics. While they have many advantages, a common cause of consumer dissatisfaction is the amount of product remaining trapped inside the pack. This 'Easy Squeeze' tube neatly addresses this problem. The steps in the shoulder weaken the structure, so the user can easily press it to squeeze out all the product. This drawing also shows the patented striping mechanism for dispensing two different formulations, which remain separate right to the last 'squeeze'.

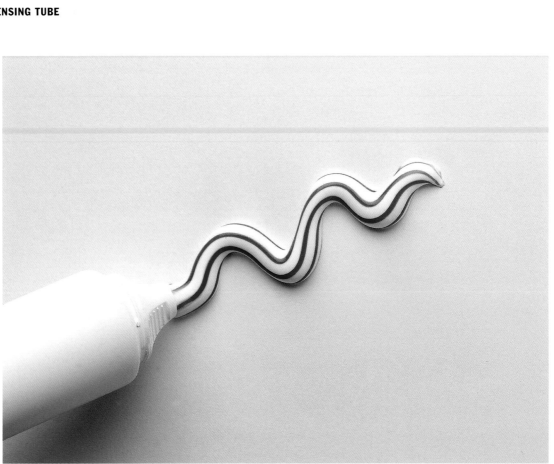

Betts UK Ltd and Unilever

132 **PLASTIC COLLAPSIBLE DISPENSING TUBE**

This neat one-piece pouring closure has a push-fit outer cap which flips up to reveal a tamper-evident membrane. To gain access to the product the membrane must be torn away by pulling up a plastic ring. The product can then be dispensed in a controlled manner through the opening via the directional pouring lip, which means a clean operation, with excess product draining back into the container.

POURING CLOSURE WITH TAMPER-EVIDENT MEMBRANE 133

Depilatory creams are traditionally supplied with a separate spatula for applying the product to the skin. This two-component design incorporates a ribbed applicator head in the snap-on closure, which allows the product to be cleanly and evenly dispensed and applied. An overcap seals off the orifice in the applicator and provides a flat plane on which to stand the pack.

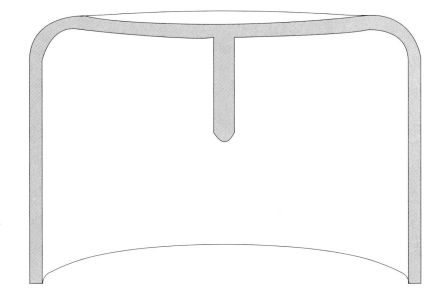

134 **CREAM APPLICATOR CLOSURE**

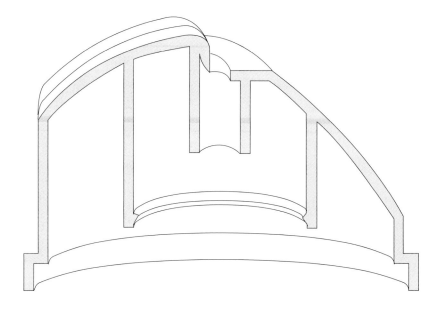

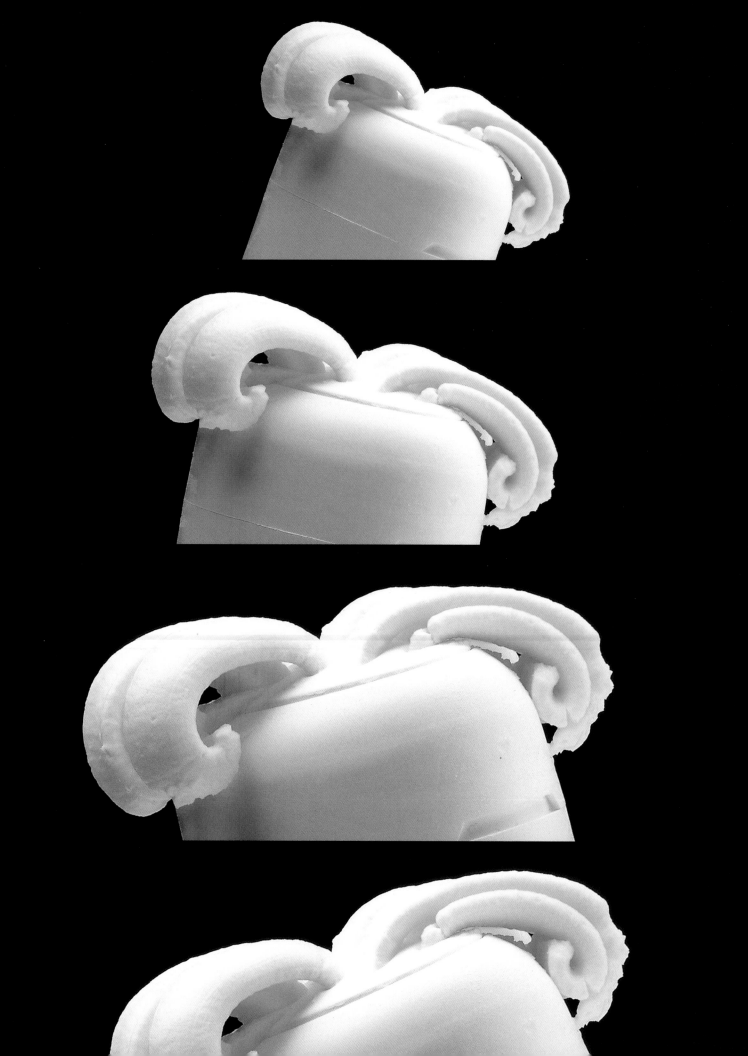

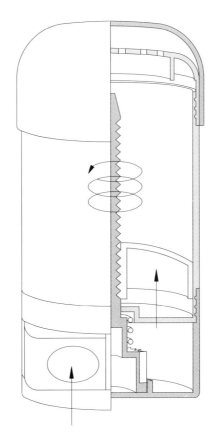

Martin Bucknell at Tin Horse, Ron Kutepof/a and Steph Cutter for Elida Fabergé

An innovative, no-mess mechanism for delivering a measured dose of a cream product, in this case an antiperspirant deodorant. The pack is easily held in one hand and two clicks on the push button deliver just the right amount of product through the dispensing holes. The smooth shape of the applicator allows the product to be easily applied to the skin. A snap-on overcap keeps the product clean and hygienic when not in use.

CREAM DEODORANT DISPENSER 137

This Japanese pack contains a pre-wash laundry liquid and it has a two-part closure. The main part screws onto the neck of the container and forms a bore seal. The moulding includes a series of concentric circles of plastic filaments, which act as a brush to scrape out persistent dirt from the clothes, thus enhancing the effectiveness of the product. There is a clear plastic screw-on overcap with a central nodule to seal off the aperture when the pack is not in use.

138 **CLOSURE WITH INTEGRAL PLASTIC BRISTLES**

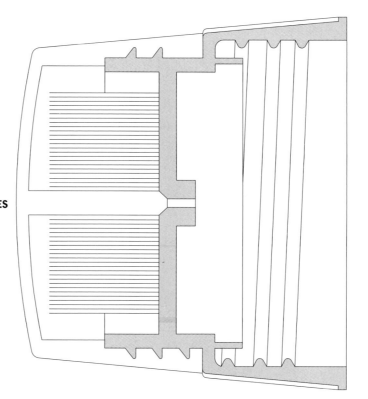

Lion Corporation, Japan
PackTrack™ CPS International Ltd.

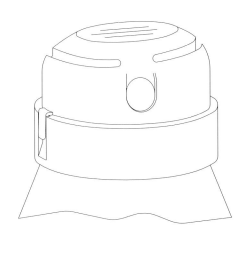

To operate this aerosol a small tab is pulled out, allowing the top section to be turned anticlockwise to a 'stop' point, to line up with the dispensing nozzle. Only then can the large actuator button be depressed to dispense the product in a fine spray. The mechanism is locked by turning the cap clockwise to its original position, thus preventing accidental depression of the actuator while the pack is in the sports bag etc.

LOCKABLE AEROSOL DISPENSING CLOSURE SYSTEM 141

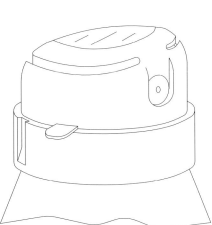

This simple one-piece screw-threaded plastic cap has a tamper-evident band which breaks and falls away from the cap on first opening. The interrupted thread allows for fast machine-application and the deep bore seal prevents leakage and maintains product quality.

142   **TAMPER-EVIDENT BORE SEAL CAP**

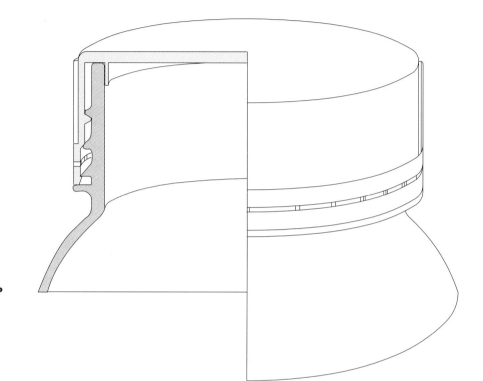

Bericap UK Ltd.

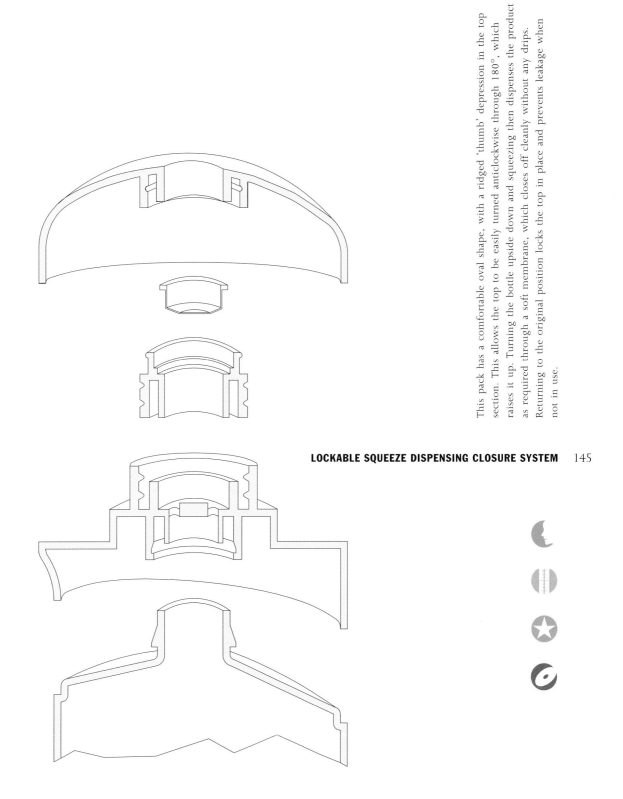

This pack has a comfortable oval shape, with a ridged 'thumb' depression in the top section. This allows the top to be easily turned anticlockwise through 180°, which raises it up. Turning the bottle upside down and squeezing then dispenses the product as required through a soft membrane, which closes off cleanly without any drips. Returning to the original position locks the top in place and prevents leakage when not in use.

LOCKABLE SQUEEZE DISPENSING CLOSURE SYSTEM 145

A dispenser with a very smooth pumping mechanism, which is suitable for dosing rich, smooth-textured creams of high viscosity. The dispenser can be readily screwed on to standard bottle neck finishes and it has a locking mechanism which prevents accidental usage or spillage. The rounded shape is soft and discreet and the pump is available with a metal or plastic collar for different aesthetic effects.

PUMP DISPENSER FOR HIGH VISCOSITY PRODUCTS

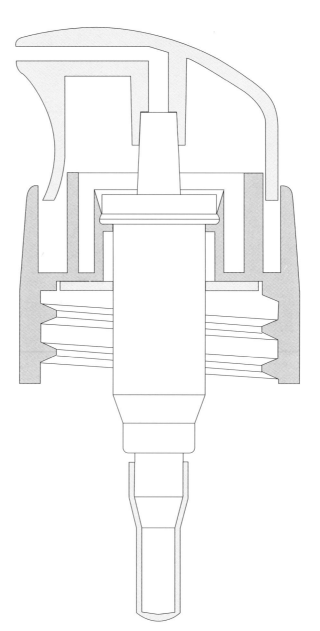

Perfect-Valois UK Ltd.

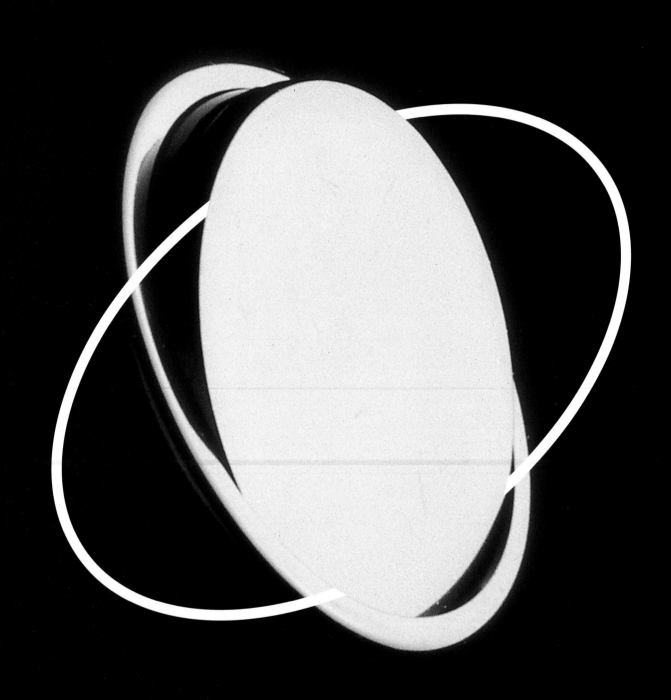

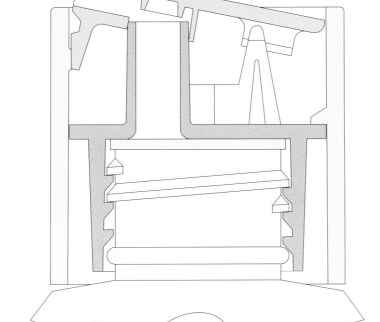

A double walled Twist Disc® cap with screw-thread to fit to standard-container-neck finishes. A gentle anticlockwise twist on the closure raises an inner disc to dispense the product. Reversing the twist locks the disc and securely closes the pack.

TWIST-OPEN DISPENSING CAP   149

The Disc Top® screw-threaded cap allows one-handed opening of a pack by pressing the top section in the area clearly identified 'PRESS' in the moulding. This causes the disc to flip up and the product can be dispensed through the nozzle. Pushing the disc down closes the pack securely. Both the opening and closing actions have an audible 'click' confirming the operation.

150 **PRESS-OPEN DISPENSING CAP**

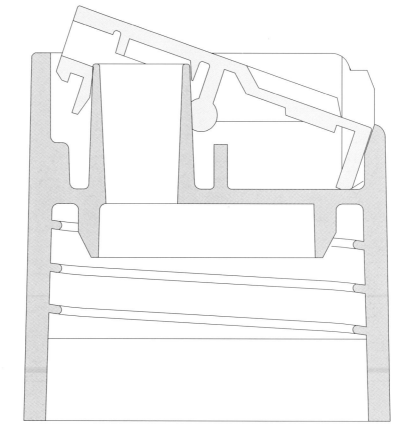

Seaquist Loffler Ltd.

Sequist Löffler Ltd.

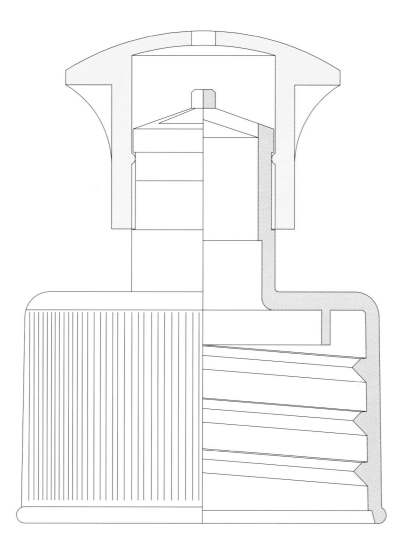

A two-piece closure system. The screw-threaded section fits to standard-container-neck finishes and has an extended open spout. The push-pull fitment snaps onto this spout and when pushed down, seals off the opening. Pulling the fitment up allows just enough clearance for the product to be dispensed by squeezing the container.

PUSH-PULL DISPENSING CAP   153

This is a well-established closure type with many industrial applications for powdered or granulated products such as sugar, cement and chemicals. The product is filled at high speed through the valve and once filled the weight of the product resting on the valve acts as a self-closing feature. An additional heat-sealing operation can be applied to the valve if required.

154 **PAPER SACK WITH VALVE CLOSURE**

Korsnäs Paper Sacks Ltd.

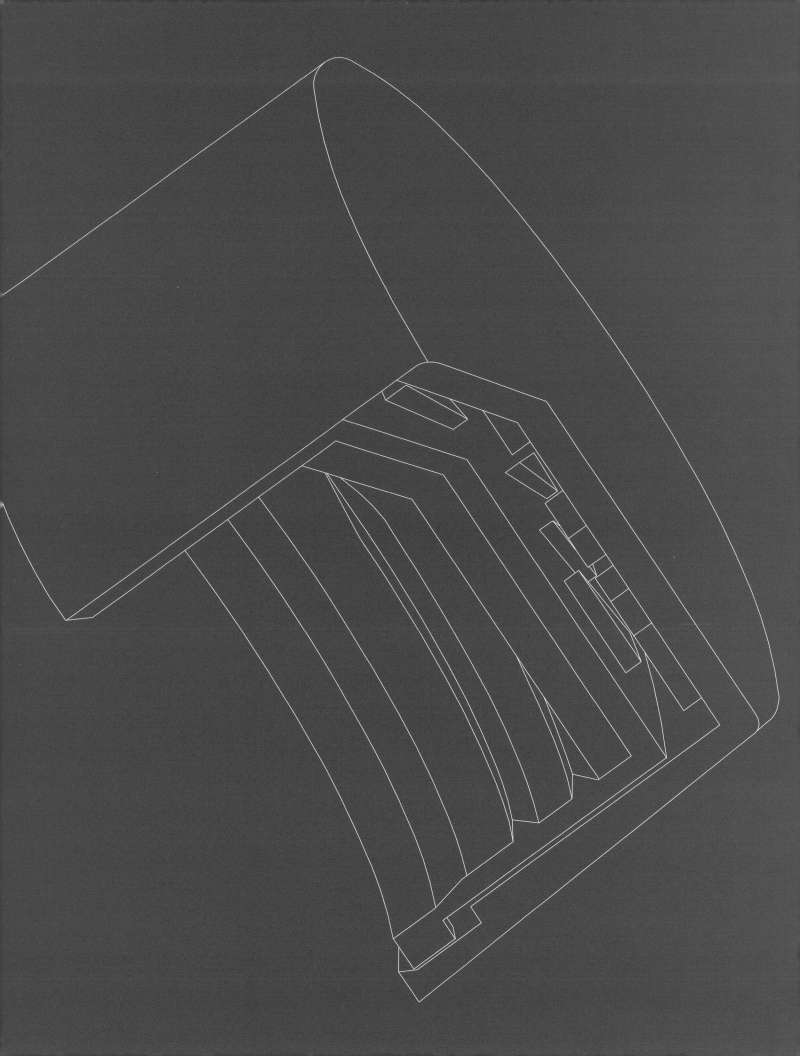

Useful addresses

Akerlund & Rausing Ltd., 26 High Street, Syston, Leicester, LE7 1GP, U.K.

Adelphi (Tubes) Ltd., Olympus House, Mill Green Road, Haywards Heath, West Sussex, RH16 1XQ, U.K.

Alcan Deutschland GmbH, Eisenwerk 30, 58840 Pletenberg, Germany.

Aquasol Ltd., Writtle College Campus, Writtle, Chelmsford, Essex, CM1 3WN, U.K.

Bel-UK Ltd., Bel House, North Court, Armstrong Road, Maidstone, Kent, ME15 6JZ, U.K.

Bericap UK Ltd., Oslo Road, Sutton Fields Industrial Estate, Hull, HU7 0YN, U.K.

Betts UK Ltd., 505 Ipswich Road, Colchester, Essex, CO4 4HE, U.K.

Boots the Chemist, 1 Thane Road, Nottingham, NG90 1BS, U.K.

Campbell Grocery Products, Hardwick Road, King's Lynn, Norfolk, PE30 4HS, U.K.

Carnaud Metalbox Food UK, Perry Wood Walk, Worcester, WR5 1EG, U.K.

Coca-Cola Great Britain, 1 Queen Caroline Street, London, W6 9HQ, U.K.

Dragon Plastics Ltd., Taffs Mead Road, Treforest Industrial Estate, Pontyprid, Mid Glamorgan, CF37 5TF, U.K.

DSS Worldwide Dispensers, Lee Road, Merton Park Estate, London, SW19 3WD, U.K.

Elida Fabergé, Coal Road, Seacroft, Leeds, West Yorkshire, LS14 2AR, U.K.

Grolsch (UK) Ltd., 137 High Street, Burton-on-Trent, Staffordshire, DE14 1JZ, U.K.

House Shokuhin Corporation, 1–5–7 Mikuriya-Sakaemachi, Higashi-Osaka, Osaka, 577–8520 Japan.

Inteac Ltd., Writtle College Campus, Writtle, Chelmsford, Essex, CM1 3WN, U.K.

ITW Fastex, Unit 12, Bilton Road, Kingsland Industrial Park, Basingstoke, Hampshire, RG24 8NJ, U.K.

Korsnäs Paper Sacks Ltd., Northfleet, Gravesend, Kent, DA11 9BZ, U.K.

Lion Corporation, 3–7 Honjo 1–Chome, Sumida-Ku, Tokyo, 130–8644 Japan.

Maddison Product Design, Walnut Tree Yard, Lower Street, Fittleworth, West Sussex, RH20 1JE, U.K.

Metal Packaging Manufacturers Association (MPMA) Ltd., Elm House, 19 Elmshott Lane, Cippenham, Slough, Berkshire, SL1 5QS, U.K.

Morplan Ltd., P.O. Box 54, Harlow, Essex, CM20 2TS, U.K.

Nestlé UK Ltd., St. Georges Street, Park Lane, Croydon, Surrey, CR9 1NR, U.K.

New Covent Garden Soup Company Ltd., 35 Hythe Road, London, NW10 6RS, U.K.

The Robert Opie Collection, Museum of Advertising and Packaging, Gloucester Docks, Gloucester, GL1 2EH, U.K.

P. I. Design Intl., 1–5 Colville Mews, Lonsdale Road, London, W11 2AR, U.K.

PP Payne, Giltway, Giltbrook, Nottingham, NG16 2GT, U.K.

Peerless Plastics Packaging, Mariner, Lichfield Road Industrial Estate, Tamworth, Staffordshire, B79 7UL, U.K.

Perfect-Valois UK Ltd., 11 Holdom Avenue, Denbigh East, Bletchley, Milton Keynes, MK1 1QU, U.K.

Ranks Hovis McDougall Ltd., The Lord Rank Centre, Lincoln Road, High Wycombe, Buckinghamshire, HP12 3QR, U.K.

Reseal-it Sweden AB, Kullagatan 6 3tr, S–252 20 Helsingborg, Sweden.

Rexam Closures & Containers, 3245 Kansas Road, Evansville, Indiana 47725, U.S.A.

Rexam Combibloc Ltd., Blackthorn Way, Houghton Le Spring, Tyne and Wear, DH4 6JN, U.K.

Rowse Honey Ltd., Moreton Avenue, Wallingford, Oxfordshire, OX10 9DE, U.K.

RPC Containers Ltd., Gallamore Lane, Market Rasen, Lincolnshire, LN8 3HZ, U.K.

Seaquist-Löffler Ltd., 5 Bruntcliffe Avenue, Morley, Leeds, Yorkshire, LS27 0LL, U.K.

Sonoco Consumer Products Ltd., Stokes Street, Clayton, Manchester, M11 4QX, U.K.

Supreme Plastics Ltd., 300 Regents Park Road, London, N3 2TL, U.K.

Techpack UK Ltd., Unit 2, Aultmore, Kingswood Road, Tunbridge Wells, Kent, U.K.

Tetra Pak Ltd., 1 Longwalk Road, Stockley Park, Uxbridge, Middlesex, UB11 1DL, U.K.

United Closures and Plastics, 1 Steuart Road, Bridge of Allan, Stirling, FK9 4JG, U.K.

West Pharmaceutical Services, Bucklers Lane, Holmbush, St. Austell, Cornwall, PL25 3JL, U.K.

Zip Pack Packaging Technology Ltd., Unit 17, Shaw Wood Business Park, Doncaster, DN2 5TB, U.K.

The PSAG

Victor International Plastics Ltd.: contact Jo Minion, Langley Road South, Salford, Manchester, M6 6SN, U.K. Tel: +44 (0)161 737 1717, fax: +44 (0)161 737 3611, Email: jo.minion@victor.mahanna.com, website: www.mahanna.com. Victor International Plastics Ltd. is a business unit of MA Hanna

RPC Containers Ltd.: contact Corrine Lawrence, Unit 1, Higham Ferrers Bypass, Higham Ferrers, Rushden, Northamptonshire, NN10 8RP, U.K. Tel: +44 (0)1933 411221, fax: +44 (0)1933 414817, Email: corinne.lawrence@rpc-rushden.co.uk, website: www.rpc-containers.co.uk

Corus Packaging Plus: contact Sue Shaw, P. O. Box 18, Ebbw Vale, Gwent, NP23 6YL, Wales, U.K. Tel: +44 (0)1495 334488, fax: +44 (0)1495 350988, Email: Sue.Shaw@britishsteel.co.uk, website: www.corusgroup.com

The Packaging Development Company: contact Barry Jones, 29 London Road, Bromley, BR1 1DG, U.K. Tel: +44 (0)208 402 2001, fax: +44 (0)208 402 2110, Email: barry@the-pdc.co.uk, website: www.psag.co.uk

Decorative Sleeves Ltd.: contact Jo Bowett, Rollesby Road, Hardwick Industrial Estate, Kings Lynn, PE30 4LS, U.K. Tel: +44 (0)1553 769319, fax: +44 (0)1553 767097, Email: jbowett@decspo.ccmail.compuserve.com, website: www.decorativesleeves.co.uk

Merck Ltd., Pigments Division: contact Mark Aartsen, Merck House, Poole, Dorset, BH15 1TD, U.K. Tel: +44 (0)1202 664727, fax: +44 (0)1202 666530, Email: mark.aartsen@merck-ltd.co.uk, website: www.merck-ltd.co.uk

Gilchrist Bros. Ltd.: contact Pete Shaw, Ring Road, West Park, Leeds, LS16 6RA, U.K. Tel: +44 (0)113 288 3200, fax: +44 (0)113 275 1690, Email: pshaw@gilchrist.co.uk

Tag Labels Ltd.: contact Gareth Farmer, 14–15 Buckingham Square, Wickford, Essex, SS11 8YQ, U.K. Tel: +44 (0)1268 767690, fax: +44 (0)1268 764774, Email: gareth@taglabels.co.uk

Useful references

International (ISO) and British Standards (BS) for container dimensions and tolerances are available for reference in many public libraries.

Consult the BSI website at www.bsi-org.uk for more details and to order copies of Standards.

The Institute of Packaging is the professional association for those involved in packaging and provides a range of useful resources, including training courses and a Packaging Bookshop.

See the IOP website at www.iop.co.uk for details.

Pira International is a packaging research organisation, with research, testing and training facilities, as well as an information centre.

See www.pira.co.uk for details.

There are a number of packaging and related packaging user journals which provide regular reviews of packaging materials and processing developments.

These include:

Packaging Magazine, tel: +44 (0)1732 364422

Packaging News, tel: +44 (0)208 901 2921

Packaging Today, tel: +44 (0)207 466 9106

Paper Focus, tel: +44 (0)1923 261555

Modern Plastics, (USA) tel: +1 609 426 7070
See website www.modplas.com

International Bottler and Packer, tel: +44 (0)1256 764180

REFERENCE BOOKS

Briston J, *Advances in Plastics Packaging Technology*. Pira International, 1992.

Emblem A, Emblem H, (Eds), *Fundamentals of Packaging Technology*. The Institute of Packaging, 1999.

Stewart B, *Packaging Design Strategy*. Pira International, 1994.

Stewart B, *Packaging as a Marketing Tool*. Pira International, 1995.

Turner T A, *Canmaking*. Blackie, 1998.

158

INDEX